This Book Belongs To:

_____

CHING AND HIS THREE TURTLES

NUMBER STORIES OF LONG AGO

BY

DAVID EUGENE SMITH

Merchant Books

PREFACE NUMBER ONE

JUST BETWEEN US, AND WORTH READING

These are the stories that were really told in the crisp autumn evenings, the Story-Teller sitting by the fire that burned in the great fireplace in the cottage by the sea. These are the stories as he told them to the Tease and the rest of the circle of friends known as the Crowd. Sitting by the fire and listening to the stories, in the lights and shadows of the dancing flames they could see the forms of Ching and Lugal and all the rest with their curious dress of long ago.

Night after night he told these tales of the ages past, stories unlike the make-believes they had often heard, stories of what might really have happened when the world was young, stories that the Crowd said were "different" because they told of much that was new, much that was curious, and much that was interesting.

So the Crowd learned many strange things that have happened in Number Land, but they learned much more than this; for the Story-Teller told them much that was interesting about the way in which boys and girls used to write in centuries long past — how Ching wrote on palm leaves, and Lugal on bricks, and Hippias on parchment. He also told them about many of the number puzzles that have delighted boys and girls for thousands of years, so that the Tease found new tricks to play on all her friends, and the Crowd found much to think about as the stories were related by the great log fire.

And you who read these stories should imagine yourselves sitting by the great log fire and listening to the Story-Teller. You should seem to see in the flames and the shadows the moving pictures of those who played their parts in Number Land when the world was learning as you do.

Is this history? Never mind. What is history but a story, and is not every story a history of something? Why bother our heads over history? For us the story is the important thing.

PREFACE NUMBER TWO

FOR THE GROWN-UPS, AND NOT WORTH READING

The story of our numbers, of the world's attempts to count, of the many experiments in writing numerals, and of the difficulties encountered through the ages in performing our everyday computations — all this is so interwoven with the history of humanity as to have an interest for every thinking person. As the world has grown, so the work with numbers has grown; when the world has faced the mysteries of the universe, numbers have assisted in solving its problems; when commerce and science have shown new needs in computation, arithmetic has always been ready to lend a hand. The history of mathematics is no small part of the history of civilization.

This being the case, it seems proper to relate at least some portion of the story of numbers to the pupils in our schools. It can be made quite as interesting as any other story of civilization, for it touches upon a subject with which the pupils in our schools are in daily contact, adding new values to the problems of arithmetic and giving a new perspective to the whole study of mathematics.

This book is intended for supplementary reading in the elementary school. It is written in nontechnical language, and the effort has been made to connect with the history enough of the human element to make it more interesting than any mere recital of facts. With it there is also joined something of the history of writing materials, this being connected naturally with the story of our numbers. Chapters I–VIII can easily be read aloud, and the Question Box at the end of each chapter can be used as a basis for conversation or for written work.

The facts stated in the book are as nearly exact as the circumstances permit. It is not to be expected, however, that changes in the form of various numerals will be considered. Such changes are of no moment in a work of this nature and do not contradict the statement that the historical facts are presented with substantial accuracy.

It is the author's hope that this little series of human incidents will create a new interest not merely in the study of arithmetic but in the story of the development of our civilization.

DAVID EUGENE SMITH

CONTENTS

There is no knowledge that is not power. — EMERSON

There is no Past, so long as books shall live. — BULWER-LYTTON

They who lived in history only seemed to walk the earth again. — LONGFELLOW

Do not then train boys to learning by force and harshness; but direct them to it by what amuses their minds, so that you may be the better able to discover with accuracy the peculiar bent of the genius of each. — PLATO

AN-AM, BEL, AND THEIR SHEEP

NUMBER STORIES OF LONG AGO

CHAPTER I

HOW CHING AND AN-AM AND MENES COUNTED

The logs are burning in the great stone fireplace in the cottage by the sea. The Story-Teller sits in his easy-chair looking at a book of curious pictures and still more curious letters. She of the teasing ways is dancing through the open door, and with her are the others who make up what she calls the Crowd, tired with the hours of play upon the beach.

"Just one little story before bedtime," says the Tease.

"Just one," chime in the others.

"Not a single word," says he of the book with the curious pages.

"Oh, just one," says the Tease.

" Just one," begs the Crowd.

" Well, just one," says the Story-Teller, who knew all the time that he would submit. " Take your chairs, then, put a new log on the fire, and listen to the story of Ching and An-am (än äm) and Menes (mē'nēz)."

The logs burned, the Crowd sat by the fire, and he of the curious book told this story:

It is so very, very long ago that not even the wisest men of China can tell the year or the century in which little Ching, the king's oldest son, played in the forests at the foot of Mount Yu, and painted a face on the shell of his biggest turtle, and told the soldier who guarded him what a lot of turtles he had. To be sure, Ching had only three turtles, but he did n't know a word for " three," and the soldier did n't, and not even the king could do more than say, "Yes, there are a lot of turtles."

For all this was so long ago that even in the oldest parts of the earth, of which China was one, most people could not count. It was before kings had palaces or crowns or royal robes, and when they were little more than savages. So we do not wonder

2

that Ching, even though he was the son of a king, could only count " one, two," everything beyond that being a "lot." This was as far as people needed to count when Ching was playing in the forest at the foot of Mount Yu, for money was not invented, and we use our numbers to-day chiefly in buying the things we need. But in those days kings had many slaves and made them work, and sent them to kill animals, and made them bring back skins for clothing and meat for food. Few people needed to count, and only these few ever learned. Even the wisest men did not know much about the numbers that we use every day, because they had no need to do so.

At the time that Ching was growing up in China there lived on the plains of Mesopotamia (měs'ŏ pŏ tā'mĭ à), in southern Asia, a boy named An-am. He was the son of Bel, a shepherd of the country afterwards called the land of Babylon. Bel tended the sheep, and drove away the wolves that prowled about at night, and kept a careful watch to see that not one of his flock should wander away. One day Bel called out to An-am, " There

3

are many sheep out there; drive them back." But really there were only a few sheep, for neither An-am nor Bel could count beyond three, and all larger numbers were called "many." Nevertheless Bel and An-am knew the sheep so well that they could tell if one was missing, just as a good shepherd dog to-day knows if one of his flock has gone astray. So An-am and Bel could count "one, two, three, many," and that was all they needed to know about arithmetic.

While Ching was playing in the forest at the foot of Mount Yu, and An-am was helping to watch the flocks that fed near the Euphrates (ū frā'tēz), another boy was living on the banks of the Nile in ancient Egypt. This boy's name was Menes, and he lived not far from the place where now the enormous dam holds back the waters of the great river. The little hut in which Menes lived was the grandest house that he or his father or his mother ever saw, and yet it had only a single room, and this was smaller than the schoolroom in which you study arithmetic.

For this was thousands of years ago, long before people had real houses, long before

4

anyone knew how to read or write, long before the world had learned how to weave fine cloth, and long before men knew any other way to make a light than to rub two pieces of wood together until one of them was set on fire. Menes was proud of what he thought was the magnificent house in which he lived, although it was only a little hut, and he was glad to be able to say, "We have a great many palm trees about our house," although there were only six. For Menes had heard his father and mother speak of one tree, of two trees, of three trees, and of four trees, but beyond that they simply said "a great many trees," for they had names for numbers only up to four, and all beyond that was a great many, just as we might speak of a great many apples.

When Ching and An-am and Menes grew to be men, and Ching became a king, and An-am became a manager of the Babylonian king's estates, and Menes became a great captain in the wars against the savages who lived in the south, Ching could only count to two, and An-am to three, and Menes to four, because this was as far as people in their countries could count in the days when the world was only just coming out of savagery, when money was unknown, and when no one measured land or buildings or the things which they traded with one another.

But many hundreds of years later other boys played in the forest at the foot of Mount Yu, and they counted "one, two, two and one, two twos, two twos and one, a lot." The world was growing, and people needed larger numbers, and so they counted as far as "two twos and one," which we call "five," and all beyond that was simply called a "lot."

And other boys helped to tend the flocks of Babylon, and their fathers taught them to count by threes, — "one, two, three, three and one, three and two, two threes, two threes and one, two threes and two, three threes, three threes and one, three threes and two, many," for they did not know a word for four, so they could n't say "four threes," and they just said "many"; but of course they said another word, using the language of ancient Babylon. The world of Mesopotamia was growing older, and people needed more

6

number names; but they still had no money, and a few such names were quite enough.

While the boys were counting to "two twos and one" in China and to "three threes and two" in Mesopotamia, Egyptian boys played under the palms where long before their time Menes had looked with pride upon his father's hut. No longer, however, was there just a hut with a single room, for the world was growing still older, and the descendants of the Menes of long ago had now a house with two rooms, and the Menes of this time had learned a new way of counting. The people along the Nile had found that the fingers of one hand would help them with their numbers, and so they made new names as far as five, and Menes now counted "one, two, three, four, five, five and one, five and two," and so on to "five fives and four"; and then he gave up and said "a great many." He could count farther than the Chings and the An-ams, but "five fives and four" is only twenty-nine, and this does not seem very far to us. But this was long before people could read and write, when they used knives made of stone, and when they thought the world was growing old, while to us it seems to have been very young.

Hundreds of years again went by, and still new Chings and An-ams and Meneses played in the forests of Yu, or on the plains of Mesopotamia, or on the banks of the Nile; but now the world began to feel that "five fives and four" was not large enough, even in ancient Egypt. Then it was that someone thought that if people could count to five on one hand, they might as well count to ten on two hands, and so the Ching and An-am and Menes of that day counted the trees and sheep by learning number names to ten, and then saying "one and ten, two and ten, three and ten," and so on to "ten tens, ten tens and one," and as much farther as they wished to go. The world had discovered that its ten fingers were useful in counting, and so it learned to count by tens; and this was one of the greatest discoveries that the world ever made. Although boys and girls speak different languages, they all have ten fingers, and so all civilized people to-day count by tens.

Near the equator, where the climate is

hot, and where people did not wear shoes, they counted their toes as well as their fingers, learning separate number names to twenty, — not "one and ten," "two and ten," and "three-ten" (thir-teen), but "eleven," "twelve," and so on, with special names, to twenty, which they sometimes called "man finished," and beyond that they counted by twenties. Some of these people wandered to other countries and carried along with them their way of counting. But most of the people of the world did as the children of Ching and An-am and Menes did, — they counted by tens. When we hear of "three-score years and ten," and when the French speak of "four twenties" instead of eighty, we have two remaining bits of the old way of counting by twenties.

Thus the world learned from Ching and An-am and Menes, and from their children and their children's children, and so on for hundreds and hundreds of years, first to count by twos or threes, and then by fives, and then by tens, and sometimes by twenties. A few people tried to count by twelves, and so we have twelve inches in a foot, twelve

9

ounces in the ancient pound, and twelve things in a dozen, but the reason why the world came to count by tens was because Ching and An-am and Menes and you and I have just ten fingers on our two hands.

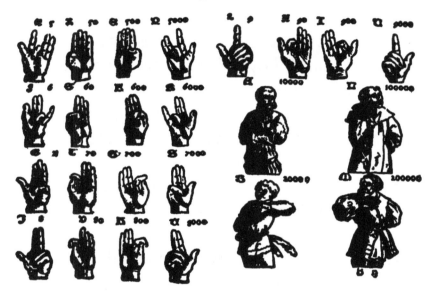

REPRESENTING NUMBERS BY THE HAND

From a book printed nearly four hundred years ago. It shows the way in which numbers were represented by the hand

Long after the early days of which we have been speaking, the world learned how to write numbers. Because different races wrote them in different ways, the merchants who traded with others whose language they did not speak represented numbers by their fingers. For at least two thousand years the merchants of different countries made number signs with their fingers in bargaining at the great fairs where they met to buy and sell the goods that thus went from country to country, — spices from India, silks from the land of Ching, wool from the ancient home of An-am, and dates from the palm trees under which Menes played many centuries before.

"Did they really count like this?" asked the Tease.

"Really," replied he of the curious book.

"I think it is funny," said the Tease.

"No funnier than your way would seem to Ching," said the Story-Teller.

"Do we have another story to-morrow night?" asked one of the Crowd.

"Not another story ever," replied the Story-Teller, "unless you go to bed."

"And then?"

"To-morrow we must all fill a question box, and each must answer every question."

"And then?" asked the Tease.

"And then? Well, then we shall see," replied the Story-Teller.

THE QUESTION BOX

1. How far could the first Ching mentioned in the story count?

2. Why did Ching not learn to count as we do?

3. What need have we for counting that Ching did not have?

4. Where did Ching live?

5. How far could the first An-am count?

6. If An-am could count no farther, how could he tell if one sheep was missing out of twenty sheep?

7. Why was it unnecessary for An-am to count farther?

8. Where did An-am live?

9. What name did the first Menes give for numbers beyond four?

10. Where did Menes live?

11. How did people come to count by tens?

12. By what other numbers than ten did the world sometimes count?

13. Why did people use one of the numbers referred to in Question 12?

14. What practical use did the world make of representing numbers on the fingers?

AHMES AND THE PRIEST
AT THE TEMPLE GATE

CHAPTER II

HOW AHMES AND LUGAL AND CHANG
WROTE THEIR NUMBERS

" Just one little story," whispered the Tease.

" Not a single word," said the Story-Teller.

" We filled the Question Box," said Maude.

" Just one," pleaded Gertrude.

" I will put on a big log," said Charles.

" And we will all keep still," said Fanny.

" Well, then, just one, but it must be short," remarked he of the curious book.

" What is the story about ? " asked George.

" This," replied the Story-Teller, " is to be about Ahmes (ä′mĕs) and Lugal (lōō′găl) and Chang (chäng) and the way in which they wrote their numbers."

This, then, is what the Story-Teller related about these boys of long ago:

It was many a long century after the last of the little Chings had learned to count by tens that there lived on the banks of the Hwangho (hwänghō′), or Yellow River, a potter whose plates and cups were known all over Shantung (shän′tŏong′) as the best to be found anywhere in that great province.

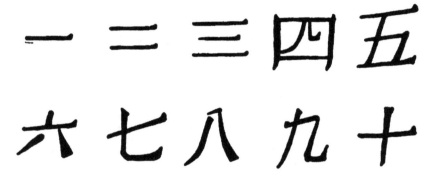

CHINESE FIGURES

Modern Chinese figures for the numbers from 1 to 10. These developed from those used by Chang

One day Chang watched his father write with a brush on a palm leaf some marks which showed how many cups there were upon the top shelf in his workshop.

Chang had learned to count the cups, but he could write ten only by making ten marks. So he said to his father, " I want to learn how to write the numbers as you do."

14

Then his father took the brush and wet it, rubbed it on a small cake of black paint, and painted some strange figures on a piece of palm leaf. These figures did not look at all like ours, for Chang's father did not know how to write numbers as we write them, and, indeed, no one in all the world at that time knew anything whatever about them.

CHINESE NUMBER

The number 789 in modern Chinese. The symbols mean 7 hundred 8 ten 9. The number is commonly written from the top down instead of from left to right

While Chang was learning the Chinese figures on the banks of the Yellow River, another boy was learning to write numbers in a far-distant country. This boy's name was Lugal, and his father was a merchant in Mesopotamia, not far from the place where An-am had lived so many years before.

Lugal's father did not use a brush and a palm leaf as Chang's father did. He took a piece of damp clay, just as you may have

15

done when you made mud pies as children, and in this damp clay he wrote his numbers.

All this was still long before money was invented and long before paper was made. The few people who could write at all made letters by pressing with a pointed stick on pieces of damp clay, the clay being then dried in the sun.

BABYLONIAN NUMERALS

Babylonian figures for the numbers from one to ten as they appear on the ancient clay tablets

Because the signs looked like little wedges they were called cuneiform (kŭ nē′ĭ fôrm) signs, a word meaning "wedge-shaped."

This does not seem to us a very good way to write numbers, but it was the only way that Lugal knew. There was no book about arithmetic in those days, and boys and girls worked hard all day in the fields or shops instead of going to school.

While Chang and Lugal were learning their curious ways of writing numbers in China and Babylon, there lived in the valley of the Nile a boy named Ahmes. Near his home was a large temple, and on the walls were many curious signs that puzzled him.

EGYPTIAN NUMERALS

Numerals from the walls of a temple at Luxor

One day when Ahmes saw a priest looking at the temple wall he asked him what the curious signs meant.

The priest replied, " These are figures stating the number of days in the king's last war."

Then Ahmes said he wished that he knew how to write numbers, and the priest told him that he would teach him how to do so if he would get some papyrus (pà pī'rŭs).

ANCIENT EGYPTIAN PAPYRUS

From a papyrus roll, showing the ancient manner of writing and illustrating. (From Breasted's " Ancient Times ")

Ahmes went home and asked his father to tell him what papryus was, and in this way he learned about a kind of paper on which the priests of Egypt used to write. In early times they had written on stone, brick, or pieces of pottery, but as the centuries went on they had invented something better.

18

There was a water plant growing in Egypt that looked like the bulrush in our own swamps, only it was much larger and taller. People found that if they cut this plant in thin strips, laid these strips close together, placed another layer crosswise over the first one, pressed these all down, and then allowed them to dry, they would have something on which they could write. This substance looked like our coarse brown paper. The water plant was the papyrus reed, and so the substance was called papyrus, and this is the word from which our word "paper" comes.

The papyrus was made in long strips, often about ten inches wide. On these strips the Egyptians wrote their books. They then rolled up the papyrus as we roll up a strip of wall paper. This is the reason that we speak of papyrus rolls instead of papyrus books. When you speak of the attendance roll at school you are simply using this ancient name.

Ahmes asked his father for a piece of papyrus and a brush and some black ink, and with these he went to the temple gate the next morning. The priest then showed Ahmes how to write numbers, and said to him, "If I should write twenty-seven thou-sand five hundred twenty-nine, I should write a number larger than anyone in the world would ever need to use." The priest thought this was true, for the world was still young, no one knew about money, and no one thought that large numbers would ever be needed.

EGYPTIAN NUMBER

The number 27,529 as the priest wrote it for Ahmes. The order and shape of the figures vary greatly in different inscriptions

Ahmes lived nearly four thousand years ago, and although he grew up and became a great man, and wrote the oldest book on mathematics that now exists, he never needed to use large numbers. The world was still too young to know much about arithmetic.

"Really, couldn't people multiply and divide numbers then?" asked the Tease.

"The answer to that question is another story," said the Story-Teller.

"Then there is another story to-morrow night?" asked Maude.

"That depends on the Question Box," replied he of the curious book.

"And then what happens?" asked Herbert.

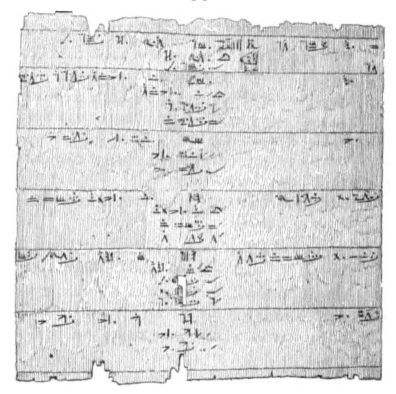

ANCIENT BOOK ON ARITHMETIC

From the oldest book on mathematics in the world, a papyrus roll written by Ahmes about 1700 B.C.

"There isn't any 'then,'" replied the Story-Teller, mysteriously. "What we call 'then' passed away long ago. There is only a 'now,' and now is the time for bed."

THE QUESTION BOX

1. Why did Chang write such strange numbers as he did instead of writing numbers as we do?

2. Why did Lugal not use a zero and thus write a number like seventy with two figures as we do, instead of writing seven times the sign for ten?

3. What materials did Chang and Lugal use for writing? Why did they not use paper?

4. What materials did the priest use when he explained the numerals to Ahmes?

5. How was papyrus made and what did it look like?

6. From what word does our word " paper " come?

7. When you speak of calling the roll, what does the word " roll " suggest to you?

8. What kind of a book did Ahmes write when he grew up and became a great man?

9. In what way are our numerals better than those of Chang and Lugal and Ahmes?

10. Could you add, subtract, multiply, and divide with the numerals of Chang and Lugal and Ahmes? What difficulties would you meet if you tried to do so?

11. Why do we call the numerals used by Lugal by a name meaning wedge-shaped?

12. Which numerals seem to you the easiest, those of Chang, those of Lugal, or those of Ahmes?

HIPPIAS
ON THE ACROPOLIS OF ATHENS

CHAPTER III

HOW HIPPIAS AND DANIEL AND TITUS
WROTE THEIR NUMBERS

"What is the story to-night?" asked the Tease as she came into the long room and stood before the fire, while the Crowd drew up the chairs.

"Story? Who said there was to be any story at all?" asked he of the curious book as he turned a new page.

"We always have a story," replied the Tease. "We have n't missed a single evening since we began."

"But we began only two nights ago."

"Yes, and this will make the third story," said George.

"But we must stop sometime," replied the Story-Teller, "and this is a good place."

"Oh, there is a great deal more that we want to know," said the Tease.

"What is it you wish to know?"

"The thing you are going to tell us," answered the Tease.

"Then," said the Story-Teller, "it must be about Hippias (hĭp'ĭ ás) and Daniel and Titus (tī'tŭs)."

And this is the story he told:

Many years after Chang had learned to write numbers in his home on the banks of the Yellow River, and Lugal to do so in

ANCIENT COINS

Coins found in Asia Minor. They are among the earliest known, dating from about 550 B.C. (From Breasted's "Ancient Times")

Mesopotamia, and Ahmes to do so by the temple on the Nile, there lived in Greece a boy who was known as Hippias.

The world was now getting old enough to have money for use in the shops, so that merchants not only traded their wares as they did in the days of Chang and Lugal and Ahmes but they sold them for copper and silver coins. This is the reason why there was more need for numbers than in the

24

centuries before Hippias played about the Acropolis (à krŏp′ŏ lĭs) at Athens and learned how the merchants wrote their numbers on a parchment roll. For not only had the people invented new ways of writing numbers but they had invented something new on which to write. They tried for a time to use long strips of leather sewed together and rolled up, and on these they wrote with a brush dipped in black ink. They then found that they could whiten and toughen the skins of sheep and calves so as to make them better suited for writing. This was first done in a city called Pergamon (pûr′gà mŏn), in Asia Minor, and from the name of this city comes the word "parchment." It was many centuries after Hippias lived before the world began to use paper.

PARCHMENT ROLL

It was on such rolls that people wrote in the time of Hippias

Hippias learned to write his numbers on parchment, using Greek characters that were very different from our numerals.

If Hippias wished to write a number like 2977, he had to use fifteen Greek letters, so you see that arithmetic must have been much harder for the Greek boys than it is for us.

I Γ Δ H X

GREEK NUMERALS

The Greek numerals for 1, 5, 10, 100, and 1000 as Hippias wrote them on parchment

As the years went on there came a time when people felt the need of some easier way of writing numbers for use in the shops of Athens. So it came to pass that another little Hippias, not long before the days when

XXΓᴴHHHHᴾΔΔΓII

NUMBER WRITTEN IN GREEK

This is the way that Hippias wrote the number 2977

Paul preached at Athens nearly two thousand years ago, wrote on his parchment roll the letters of the Greek alphabet to represent numbers. The Greek name for the first letter was *alpha* (ăl'fà), and the second letter had the name *beta* (bā′tà). When Hippias learned his A B C's he learned his *alpha-betas*, and from this name we get our word "alphabet."

While Hippias was learning to write numbers in Athens, a boy named Daniel was living on the slope of the Mount of Olives. This boy went daily to Jerusalem with fruit which his father sold in the market place. He

$$A' \; B' \; \Gamma' \; \Delta' \; E' \; F' \; Z' \; H' \; \Theta' \; I'$$

GREEK LETTER-NUMERALS

The first ten numbers as the Greeks represented them by letters about two thousand years ago. For larger numbers they used other letters : K′ for 20, Λ′ for 30, and so on. They placed a mark (/ or ′) by each letter to show that it stood for a number

needed to know how to write numbers, for the prices of the melons and figs were written on small boards and put upon his father's fruit stall. It was for this reason that Daniel's father taught him the smaller numbers that everyone needed to know. In those days, however, a man would work for a penny a day, and so most people had little need for numbers higher than ten, and numbers above a few thousand were rarely used by anyone.

The numerals which Daniel learned were only the first few letters of the Hebrew alphabet, just as those learned by Hippias were the first few letters of the Greek alphabet.

אבגדהוזחטי

HEBREW LETTER-NUMERALS

The first ten numerals as learned by Daniel, being the first ten letters of the Hebrew alphabet. They read from right to left, the Hebrew language being written in that way

You can see that such a way of writing numbers must have made multiplication and division very hard.

While Hippias was playing in the streets of Athens, and Daniel was carrying fruit from

EARLY ROMAN NUMERALS

These nine characters represent the numbers 1, 5, 5, 10, 50, 100, 500, 500, and 1000 as written by the early Romans. Notice the two ways of writing 5 and 500

the Mount of Olives to Jerusalem, Titus was playing about the streets of Rome and attending a school near the great forum of the city.

28

The teacher showed the boys the way in which the ancient Romans wrote their numbers and also taught them the numerals that were then used in the shops of Rome.

IVXLCDꞒꝹ ∞

LATE ROMAN NUMERALS

Number forms for 1, 5, 10, 50, 100, 500, 1000, and 1000 used by the Romans about the beginning of the Christian era. Notice the two ways of writing 1000

If Titus had wished to write a number greater than a million he would have been very much puzzled, for in those days people rarely had any need for numbers above a few hundreds

CIↃCIↃDCCLCIↃCIↃDCCCLXXXXVIIII

ROMAN NUMBER

One of several ways of writing 2,752,899 at the time when Titus lived. The Romans themselves were not uniform in writing such large numbers

or a few thousands. Titus would probably have written such a large number in words.

When we see the Roman numerals on a watch or a clock we should know that Europe used them commonly until after America was discovered, that they were of no value in multiplying and dividing, but were fairly convenient for writing the small numbers needed in everyday business affairs.

Titus liked to puzzle a chum of his named Caius (kā'yŭs), and one day he asked him this question: "What is the number that becomes one more when one is taken away from it?"

"Your head," replied Caius, "must be just plain wood."

But when Titus wrote IX on the stone pavement and said to Caius, "Now take away the I and tell me what you have left," Caius saw that the wooden head had something in it after all.

Then Caius, remarking that he could think of many other numbers that would answer just as well, asked this question: "What is the number that becomes ten more when ten is taken away?"

Titus then asked Caius if he knew that half of nine was four, and Caius replied that he must be dreaming. But Titus pointed again to IX and asked Caius to take the upper half of it and see if it was not IV. Then Caius said that he could show that half of twelve was seven.

"That is nothing," said Titus; "half of thirteen is eight."

"That is easy," said Caius; "but can you take one hundred from four hundred and have five hundred left?"

"I think I have now told you enough for to-night," said the Story-Teller. "How many of you can show that half of twelve is seven?"

"I can," said Emily.

"And I," said half a dozen others.

"How many of you can show that half of thirteen is eight?"

"We all can," said the Crowd.

"Who can show how Caius took one hundred from four hundred and had five hundred left?"

Not everyone could do this. Can you do it?

"Who can tell me some other curious puzzle about the Roman numerals?"

"We all can," said the Crowd.

"Then," said the Story-Teller, "I think you had better fill the Question Box."

"If we fill it do we have another story to-morrow night?" asked Charles.

"Well, if you don't fill it you certainly will not have the story," said the Story-Teller.

THE QUESTION BOX

1. In the days of Hippias what was used instead of paper for writing purposes?

2. Why did the invention of money make it necessary for people to know more about counting and about working with numbers?

3. Why were the numerals used by Hippias not so convenient as those which you use?

4. What does the word "alphabet" mean, and how is it connected with the expression "Learn your A B C's"?

5. Was Daniel's way of writing numbers more nearly like that of Hippias or like the one used by Titus? Why was it not so good as ours?

6. Where have you seen the Roman numerals practically used, outside of your school work?

7. Why are the Roman numerals not so good for computing purposes as ours?

8. How many ways do you know of writing the number four in Roman numerals?

9. If you could not use our numerals, which numerals that you have read about would you prefer to use?

10. Titus and Caius found some amusing things connected with their numerals. Caius said he could show that half of twelve is seven. How did he do it? Can you, in the same way, show that half of thirteen is eight?

CHAPTER IV

HOW GUPTA AND MOHAMMED AND GERBERT WROTE THEIR NUMBERS

"No Story To-Night," wrote the Story-Teller in large letters on a piece of paper, smiling as he did it. Then he pinned the notice on the edge of the shelf just above the fireplace, went back to his easy-chair, opened his curious book, and pretended to read.

A few minutes later a certain small lady tiptoed in and then tiptoed out again. There was the sound of whispering and quiet laughing in the hall, and then the door opened softly, the Tease once more tiptoed in, crossed out the " No," and with her pencil wrote above it the words " A Real Good."

Then the Crowd rushed in and all read aloud, " A real good story to-night."

33

"What is the use in arguing with the Crowd?" asked the Story-Teller.

"No use at all," answered the Crowd.

A Real Good ~~No~~ Story To-Night

THE STORY-TELLER'S NOTICE AND HOW IT WAS
CHANGED

But all this time the Story-Teller was just as anxious to begin as the Crowd was to listen. This, then, is the story he told:

It was a hot day, and Gupta (gōōp′tä), brown of skin and brown of dress, played under the shade of the bamboo on the banks of the Indus (ĭn′dŭs), the great stream that gives the name to India.

Gupta was glad to be out of the hot sun, glad to splash in the water with a branch of bamboo, glad to be living in that ancient land — ancient even when he played by the Indus two thousand years ago. Gupta had never been to school, for there were no

34

schools like ours to attend, and so he could neither read nor write, and he had little idea of Europe and none whatever of America.

Of all the people he knew, only the priests in the temple near his home could read and write, and no one in all Europe or Asia or Africa had ever heard of America —for this was many hundred years before Columbus was born.

The priests soon noticed that Gupta was brighter than the other boys in the village, and so they took him to the temple and taught him to read and write. They also taught him to write the numbers to four, which was as far as they themselves were certain how to write them, but these numbers did not look at all like most of ours.

ANCIENT HINDU NUMERALS

The first four numerals, for 1, 2, 3, and 4, as Gupta learned them more than two thousand years ago

The first three numerals were simply straight marks, like those the Romans used, and the four looked like our plus sign.

Gupta learned to write numbers with a sharp iron pencil which scratched slightly into a piece of palm leaf. Books were always

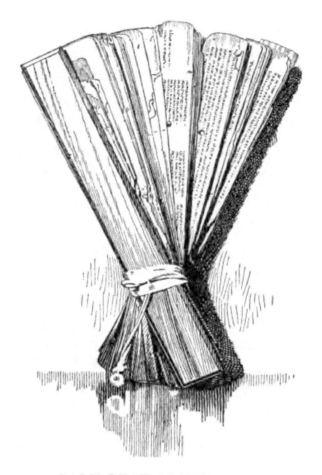

PALM-LEAF ARITHMETIC

An old arithmetic from India, written on strips made from the
leaf of a kind of palm

written in this way in India at the time that
Gupta lived, and indeed for many years there-
after, and they looked very different from the
book which you are reading.

36

When Gupta grew a little older he learned of another kind of numerals. Some priests had seen these numerals cut on the walls of

					+	⸀6	6	6	ʃ	⅄	⅄	匸
1	2	4		6	50	50	200	200	200

EARLY NUMERALS

Probable number forms from inscriptions on stone made in the third century B.C.

a cave where pilgrims often rested for the night. It is here that we have almost the

—	=	❯	❰		५	7	ʔ	α	α	α
1	2		4		6	7	9	10	10	10

O	⅂	∞	⅄	⊤	⅄	⅄	⅄	⅄
20	60	80	100	100	100	200		400

⅄	T	T	⅄	⅄	⅄	⅄
700	1000		4000	6000	10,000	20,000

EARLY NUMERALS

Probable number forms from inscriptions made in a cave in India in the second century B.C.

first traces of our present system of writing numbers, and they are more than two thousand years old. There was no zero, however;

37

no one could write a number like 207 in the way that we write it, and so the figures were no better than those of Hippias, Daniel, or Titus, or even of Lugal, Ahmes, or Chang.

Long after Gupta died, and about a thousand years ago, someone had the wisdom to invent the zero, and after that it was easy to write numbers as we write them to-day.

About the time that the zero was invented there was born in the country near the Caspian Sea a boy whose parents named him after Mohammed (mo̊ hăm'ĕd), the great religious leader of the Arabs.

Little Mohammed was a very bright boy, and Moses his father let him study with one of those wise men who watched the stars and thus told the time, for clocks and watches were then unknown.

Mohammed became so well known as a scholar that while still a young man he was called to Bagdad (băg däd') to be the caliph's (kā'lĭf) astronomer. The caliphs were the kings of the country about the Tigris (tī'grĭs) River, and many stories are told about them in the Arabian Nights Tales. These tales describe Bagdad at about the time that our

38

Mohammed lived there, and they tell of
Harun-al-Rashid (hä rōōn′ är ra̍ shēd′), a name
which means Aaron the Just. Mohammed

١ ٢ ٣ ٤ ٥ ٦ ٧ ٨ ٩ .

ARABIC NUMERALS

Numerals from 1 to 9, with a dot for zero, as used by the Arabs.
They are not like the numerals that we sometimes call Arabic

the son of Moses knew Harun's son, and the
two used to study and work together in the
caliph's observatory at Bagdad.

Mohammed the son of Moses found the
Arabs of Bagdad using numerals that were
quite different from those used by us to-day.
From certain wise men who came to Bagdad

१ २ ३ ४ ५ ६ ७ ८ ९ ०

HINDU NUMERALS

Numerals from 1 to 9, with zero, used by the Hindus and taken to
Bagdad about 1150 years ago

from India about this time, however, Mohammed learned the numerals used in that
country, and these were somewhat like ours.
He believed that they were better than the

39

ones the Arabs used, and so he wrote a book about them. This book was taken to Europe by some traveler and assisted in making known in that part of the world the numerals that we use. Because the numerals came to Europe from Arabia they were called Arabic numerals, but they were not used by the Arabs then nor have they been generally used by them since that time.

OLD EUROPEAN NUMERALS

Oldest example of our numerals known in any European manuscript. This manuscript was written in Spain in 976 A.D.

In France nearly a thousand years ago there lived a boy named Gerbert (zhĕr bâr'). He was so promising as a student that the priests whose school he attended sent him to Spain with a nobleman whom they knew, so that he might learn still more from travel. There he probably met with Arabs who knew about the Hindu numerals, because part of Spain was then under Arab rule, and when

40

he went to Rome he explained these numerals to others. He became one of the most learned men of his time and was elected pope, taking the name of Sylvester II. In this way the numerals that we often call Arabic were brought to the attention of

1	2	3	4	5	6	7	8	9	0	
										Twelfth century
										1197 A.D.
										1275 A.D.
										c. 1294 A.D.
										c. 1303 A.D.
										c. 1360 A.D.
										c. 1442 A.D.

CHANGES IN OUR NUMERALS

This shows some of the changes made in our numerals from about the time of Gerbert to the time when Europe began to print books. After printing was invented there were few changes

learned men in Europe about the year 1000, although they had been known a little before this time in Spain and possibly in Italy.

The handwriting of people changes so much from century to century and from country to country that the numerals which Gerbert used were very different from those which we study in school and which everyone in Europe and America uses to-day. Little by little, however, they came more nearly to resemble those which we now see in our arithmetics.

Printing from movable type was invented in Europe about 1450, although printing from engraved blocks was known much earlier. After the numerals were first printed they did not change very much. We use about the same figures that Columbus used. This is because printed letters and figures do not change as rapidly as written letters and figures.

An Italian boy named Leonardo (lä ổ när′dō), who was born about seven centuries ago in Pisa (pē′zä), learned about these numerals in an Arab school in northern Africa, where his father was working. When he became a man he described them in a book which he wrote, and this assisted in making them known. But of course the book was not printed then, because printing was not yet known in Europe. He left the book in manuscript, and many people read it, especially in Italy, where it was written, and in this way they learned about these numerals.

"Did all these boys really live?" asked George.

"Really; every one of them."

"What are the funny marks under the picture of Gupta?" asked Clara.

"That is the way he wrote his name in his language. People in India did not know our alphabet in those days. They had a curious one of their own."

"But the marks don't look like those under Mohammed's picture," said Helen.

"That is because Mohammed wrote his name in Arabic," said the Story-Teller.

"Gerbert was a French boy," said Edward, "so why did n't he use French letters?"

"That is because the French use Latin letters. Did you ever think how odd it seems that Americans speak English and write in Latin letters?"

"How did they ever work with such curious figures?" asked Maude.

"That is another —"

"So there is another story!" interrupted the Crowd.

"That depends on the Question Box," said the Story-Teller.

THE QUESTION BOX

1. How did Gupta learn the numerals?

2. Describe the appearance of books in Gupta's time and country.

3. Why were the numerals used by Gupta no better than the others about which you have heard? What important symbol was lacking? About when was this symbol invented?

4. When Mohammed the son of Moses went to Bagdad, did he find the Arabs using our numerals? About when did our numerals reach Bagdad?

5. What did Mohammed the son of Moses do to make our numerals known in Bagdad and also in Europe?

6. About when did Gerbert live, and how did he come to know about our numerals? What prominent position did he hold?

7. About when did Leonardo of Pisa live, and what did he do to make our numerals known?

8. Why did Leonardo not have the book printed which he wrote when he became a man?

9. About when was printing from movable type invented in Europe?

10. What kind of printing was known before the invention of movable type? Why is it better for the printers to use movable type?

ROBERT RECORD LEARNING THE USE OF COUNTERS

CHAPTER V

HOW ROBERT AND WU AND CAIUS ADDED NUMBERS

" I don't see how they ever added numbers when they wrote them as the Romans did," said Irene, just to get the Story-Teller started.

" The numerals that Hippias used were worse," said Edward.

" But Daniel's were still worse," said Will.

"Ahmes and Lugal had the worst of all," added Dorothy.

" I see," said the wise old Story-Teller, " that the Crowd has planned all this talk just to get me started "; at which the Tease smiled significantly.

" Well, it is true that they all had trouble with their numerals," he continued, "and to-night I am going to tell you a story about some of the troubles of three boys who found ways of adding numbers in spite of their awkward numerals."

When Caius was a boy he attended a kind of business school. There he learned how to write, to read the parchment rolls that told of business customs, and to perform the only two operations with numbers that were then considered absolutely necessary. These operations were addition and subtraction, and when we come to think of it they cover a large part of our business arithmetic to-day. It is not often that we need a long multiplication or a long division.

If you are asked to add 257 and 369 you find the work so easy that you can hardly imagine that it would ever trouble anyone. But when his teacher told Caius to add these numbers written in the Roman numerals, the problem did not seem so simple. It is really easier, however, than one would at first think, and if arithmetic required only adding, the Roman numerals would not be very difficult. In the first place the Romans usually wrote IIII instead of IV, VIIII instead of IX, and so on, and if you do this you will find it about as easy to add with Roman numerals as with our own.

After learning to add, which he found quite easy, Caius learned to subtract, and that was also easy; but when he came to multiplying or dividing with Roman numerals he at once found himself in a great deal of trouble.

HOW CAIUS ADDED 257 AND 369

Caius had six I's, so he wrote VI below; he had two V's, so he wrote X below; he had one X, so he wrote X below that; he had two L's, so he wrote C below; and he had five other C's, so he wrote D below them

Although Caius learned how to add and subtract with Roman numerals, there was another method that was more common. He learned that pebbles were used in working with numbers. He found that he could draw lines on a board,— one to represent units, another to represent tens, and so on,— and that he could then place pebbles on these lines.

By placing the pebbles on the lines for units, tens, and so on, he could add quite rapidly. Of course he could not add as rapidly as we

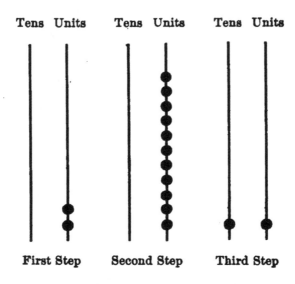

HOW CAIUS ADDED WITH PEBBLES

If Caius wished to add XXII and CXXXVIIII, he first placed 2 pebbles on the units' line, as shown in the First Step. He then placed 9 more, as shown in the Second Step. He then took away 10 of these pebbles, and added one pebble to the tens' line, as shown in the Third Step. The rest of the work is shown on page 49

do, but no one could work very quickly with numbers in those days, and Caius did as well as anyone else. People did not have much work to do with numbers, and they were not in as great a hurry as we are.

Of course Caius spoke Latin, and so he did not use the word "pebble," but used the Latin word *calculus* (kăl′kŭ lŭs), which means

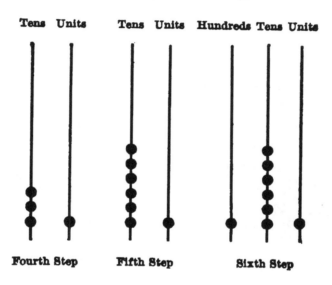

HOW CAIUS ADDED WITH PEBBLES

After the Third Step, as shown on page 48, Caius added 2 pebbles to the tens' line because of the XX in XXII, as shown in the Fourth Step. He then added 3 more because of the XXX in CXXXVIIII, as shown in the Fifth Step. Finally he drew a line for hundreds, and on this he placed one pebble because of the C in CXXXVIIII. His answer was CLXI

"pebble." So instead of saying that Caius pebbled the answer, we say that he calculated it, using a word that sounds more like Latin, and so we see what our words "calculate" and "calculation" originally meant.

49

Besides using pebbles, as he sometimes did and as was usually done in earlier times, Caius also used small circular disks like our checkers or button molds, and these he

GREEK WAX TABLET

This tablet was used by a Greek schoolboy early in the Christian era. The boy wrote part of the multiplication table upon it, and it can still be read. It is now in the British Museum

called *calculi* (kăl′kŭ lē), which is the plural of *calculus*. When he and the other boys went to school they did not carry notebooks, for paper was unknown, nor did they carry slates; but they sometimes carried small boards covered with wax, and on this wax they wrote with pointed sticks that looked somewhat like our pencils, erasing their work by simply smoothing the wax. They always carried their calculi, however, if they were studying

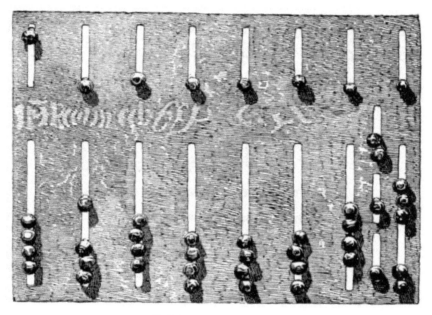

ROMAN ABACUS

Ancient bronze abacus used by the Romans. This is now in the British Museum

arithmetic. Indeed, practically everyone who did any work with numbers had at hand a small box or bag of these calculi.

Sometimes the merchants used a little calculating machine called an abacus (ăb′à kŭs), in which the calculi moved in grooves.

51

While Caius was learning to add and subtract with the Roman numerals and with calculi, a Chinese boy, whom we may know as Wu (wōo), was learning to add and subtract with the numerals which Chang had studied many years before. He too found it necessary to use something like the calculi which

ANCIENT CHINESE STICK NUMERALS

The number 1267 expressed by means of rods or sticks by the old Chinese method

Caius had worked with, but instead of pebbles or small disks he used rods made of bamboo.

Wu thought he was doing something remarkable if he added two large numbers in two minutes. You would probably add them in a few seconds, but think how much longer it would take if you had only a little pile of sticks with which to work. It took Wu some time to lay the sticks out to represent a number and still longer to represent two numbers and then to find their sum.

This plan of computing by bamboo rods was carried by the Chinese to Korea, but

KOREAN RODS

Computing rods made of bone. Until quite recently these were used in the schools of Korea. The numbers were represented somewhat as the Chinese numerals were, as shown on page 52

the Koreans used rods made of bone, and continued to do this in their schools and in business calculations until quite recently.

It was more than a thousand years after Wu learned to add with the bamboo rods that the Chinese adopted the old Roman idea of having the calculi fastened to an abacus, and so they invented their reckoning board,

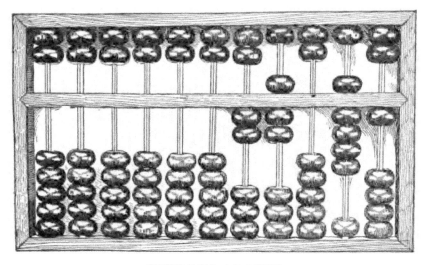

CHINESE ABACUS

Chinese abacus, or *suan pan*. This instrument is used everywhere in China to-day

or *suan pan* (swän pän). This they use in the schools, banks, and shops throughout all of China even to this day, and they often calculate more rapidly in this way than we can with pencil and paper. You have probably seen the *suan pan* used if you have ever visited a Chinese laundry in any of our cities.

54

About three hundred years ago there went to school in Japan a boy named Seki (sā kē), whose teacher had recently learned about an improvement which the Japanese had made on the Chinese abacus.

So Seki learned how to compute on the *soroban* (sō′rŏ bän′), which is the Japanese form

JAPANESE ABACUS

Japanese abacus, or *soroban*. This instrument is used everywhere in Japan to-day

of the *suan pan* of China, and this instrument is used everywhere in Japan to-day.

Seki grew up and became the greatest mathematician of Japan, and his name is known everywhere in that country and is also familiar to many mathematical scholars in other parts of the world.

Many Japanese can add and subtract more rapidly with the *soroban* than we can with pencil and paper.

55

About four hundred years ago there was born in England a boy named Robert Record (rĕk'ŏrd). When he went to school he was taught to use the Roman numerals, and he added, subtracted, multiplied, and divided numbers very much as Caius had learned to

—————————————	Ten thousands
	Five thousands
———————✕————————	Thousands
	Five hundreds
—————————————	Hundreds
	Fifties
—————————————	Tens
	Fives
—————————————	Units

PLAN OF A COMPUTING TABLE

This plan shows the arrangement of lines on the kind of computing table used in most parts of Europe in the Middle Ages. It was still extensively used when Columbus made his voyages to America

do when he computed with the aid of calculi about fifteen hundred years earlier.

Two important changes, however, had been made in that length of time. Instead of having places on the computing table for only units, tens, hundreds, and so on, the spaces between the lines were used for fives, fifties, five

56

hundreds, and similarly for all other fives, and the lines were now horizontal. We do not know when these changes were made, but they came in what are called the Dark Ages.

Because Robert cast the calculi down on the board, he spoke of "casting an account," and

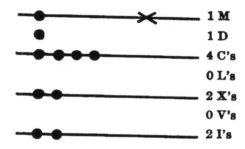

HOW ROBERT WROTE 1922 WITH COUNTERS

This shows how Robert Record represented MDCCCCXXII, or 1922, by counters. We sometimes see the number written MCMXXII, but this is not the old way of writing it. The cross was always placed on the 1000's line so as to aid the eye in reading the number. It finally suggested the use of the comma in writing a number like 47,256

this expression is sometimes used to-day. The board on which he counted was generally called a "counter." We buy goods over the counter to-day, not thinking what the word originally meant. Robert also called the calculi "counters," and if you have read Shakespeare you may have seen the expression "a counter

57

A D D I T I O N.

Master.

The easiest way in this arte, is to adde but two summes at ones togyther: how be it, you maye adde more, as I wil tel you anone. therfore whenne you wylle adde two summes, you shall fyrste set downe one of them, it forceth not whiche, and then by it draw a lyne crosse the other lynes. And afterwarde sette doune the other summe, so that that lyne maye be betwene them; as if you woulde adde 2659 to 8342, you must set your sumes as you see here.

And then if you lyst, you maye adde the one to the other in the same place, or els you may adde them bothe togyther in a newe place: which way, because it is most plynest

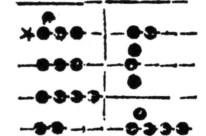

PAGE FROM ROBERT RECORD'S BOOK

This page from Robert Record's "Ground of Artes" was printed nearly four hundred years ago

caster," meaning a man who could calculate only with counters and not with pencil and paper as we do. We sometimes use counters to-day in keeping the score in games.

Robert added numbers much as Caius did, placing the counters on the lines and in the spaces. When he had five counters on a line or in a space he took them up and carried one counter to the next place. You now see how we came to say "carry one" when we add, a counter being actually carried to the next place in the days of Robert Record.

When Robert became a man he wrote several books and used his influence to have the English people give up the Roman numerals.

We are apt to smile at Robert's use of counters

A MODERN ADDING MACHINE

This kind of machine is used in banks and in many large establishments

and to think that we are much wiser than he because we use pencil and paper. But it is well to know that most of the world to-day finds it better not to use pencil and paper for computing. The Russians very generally use an abacus, many Persians use one, the Chinese use the *suan pan*, and the Japanese use the *soroban*. Even in Europe and America our most important computation is done by machinery, our banks all using adding machines, and our large corporations using machines for multiplying and dividing. Workmen in shops of various kinds often use a slide rule. Probably all of you have seen such an instrument,

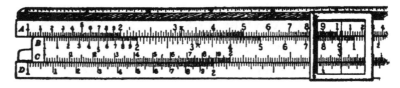

PART OF A SLIDE RULE

Such slide rules are extensively used in many industries to-day

and some of you may have learned how to multiply by sliding part of it along to the right or left.

This shows how Caius and Wu and Robert and many other boys in various parts of the world met the difficulties arising from poor numeral systems and overcame these difficulties by machinery, and how we are to-day coming to use machinery to help us in our work with numbers, especially in banks and workshops.

60

"It must have taken a long time to get an arithmetic lesson in those days," said Emily.

"No longer than it takes now," said the Story-Teller.

"How could that be?" asked Charles.

"It is just as easy as anything. All you have to do is to make the arithmetic lesson shorter, and that is just what was done."

"I wish they would do that in my school," remarked Margaret.

"Do you really?" asked the Story-Teller, "and do you want to do your work with counters on a board?"

"Well, that's different," Margaret answered, "and I think after all that our way is better than theirs."

"I don't see yet how they multiplied with the Roman numerals," said Helen.

"You are just trying to get me to promise another story," said the wily old Story-Teller. "Don't forget the Question Box. Now all of you get off to bed and don't bother me any more to-night."

"What about to-morrow night?" asked the Tease.

"To bed!" ordered the Story-Teller.

THE QUESTION BOX

1. How could Caius have added the numbers XXVII and LXXXVIII without using calculi?

2. What were the calculi and what is the origin of our word "calculate"?

3. How could Caius have added the numbers XXVII and LXXXVIII by using calculi?

4. What did Roman schoolboys take to school to help them in performing their calculations?

5. What were the wax tablets of the Greek and Roman schoolboys, and how were they used?

6. What people computed with bamboo rods in ancient times, and later with a *suan pan*?

7. In what country did the people compute with rods made of bone? Where is that country?

8. How did Robert Record compute?

9. What is the origin of our word "counter" as used in stores?

10. What nations still make much use of the abacus?

11. What kinds of machines or instruments are used for purposes of calculating in many of the most civilized countries to-day?

12. Of all the different ways of calculating that have been mentioned, which would you rather use if you were multiplying one large number by another?

CHAPTER VI

HOW CUTHBERT AND LEONARDO AND JOHANN MULTIPLIED NUMBERS

"So you wonder how people multiplied numbers, do you?" asked the Story-Teller, as the Crowd marched into the room, and George put another log on the fire, and the Tease stood wondering how it happened that the story was coming so easily. "Well, you see they just multiplied, and that's all."

"But how did they do it? That's the question," said the Tease. "How could anyone multiply with the Roman numerals, or how did anyone ever learn how to multiply, anyway?"

"When Ching was a boy no one could multiply at all," said the Story-Teller.

"I wish I could have gone to school then," said Fanny.

"There were n't any schools," said Charles.

"So much the better," remarked the Tease.

"As I have told you, those who did not go to school had to work all day, and how would you like that?" asked the Story-Teller.

The Crowd agreed that, after all, schools and multiplication and arithmetic were easier than digging and planting all day long, and so the Story-Teller continued:

"When Caius and Titus multiplied they did not use the Roman numerals; they used calculi, or counters, as they did when they added numbers. This story is too long, however, so I shall only tell you about the way in which the world learned to multiply as we do. Listen then to the story of Cuthbert (kŭth'-bẽrt) and Leonardo and Johann (yō hän')."

When Leonardo of Pisa went to school to the Moorish teacher on the northern coast of Africa, over seven hundred years ago, and learned how to write the numerals which we now use, the first great advantage that he found was in multiplying. Some of his boy friends in Pisa were probably using counters, or calculi, for this purpose, but his old Moorish schoolmaster could multiply much faster than any of them. He used the numerals that had come from the East, and so Leonardo was anxious to learn the new way.

If you have never seen a modern calculating machine, you would be interested to see how it works. You would watch the setting down of the multiplicand on certain keys, then you would see the multiplier set down, and then, on one of these modern machines, you would simply touch an electric button and the machine would make a little buzzing sound, after which you would read the answer upon it.

Some such curiosity and interest must have been Leonardo's when he first saw his old Moorish schoolmaster multiply one number of three figures by another number of three figures. With the counters this would have taken him several minutes, but his teacher did it in only one minute. Can we wonder, then, that Leonardo was anxious to learn what seemed to him to be a wonderful trick in the multiplication of numbers?

How do you think the Moorish schoolmaster proceeded? Of course we do not know exactly, but we know how Leonardo gave the work in the book which he wrote more than seven centuries ago. He first wrote the multiplicand; above that he placed the multiplier; and above that he wrote the product. This is our plan turned upside down. If you could ask Leonardo why he turned the work

```
┌─────────┐
│   392   │
│    8    │
│   49    │
└─────────┘
```

HOW LEONARDO MULTIPLIED

From Leonardo's arithmetic of 1202. It represents the multiplication of 49 by 8, the product being 392

upside down, he would say that his work was right, but ours is upside down. It all depends on how we are taught.

It was about three hundred years after Leonardo that Cuthbert Tonstall (tŭn'stăl) was born in England. In those days it was still the custom in most countries to compute by counters, but Cuthbert went to a school where they taught the new method. In 1522 he published the first arithmetic printed in England. Although this was about a hundred

66

years before the Pilgrims landed at Plymouth, it is interesting to know that only thirty-four years later, in 1556, an arithmetic was printed in the City of Mexico.

We do not know how Cuthbert was taught to multiply, but his book tells us what he thought was the best method. This was very

```
    60503
     4020
    00000
   121006
    00000
  242012
 243222060
```

HOW CUTHBERT MULTIPLIED

From Cuthbert Tonstall's arithmetic of 1522, showing the method
of multiplication used by him

much like our own except that he did not know our short way of multiplying by zero, or, as he called it, " by a circle."

Thus we see that this great man could not multiply as easily as you can.

Just before Cuthbert's time there was a boy named Johann Widman (yō hän' vĭd'män) attending an arithmetic school in Germany. In those days it was not the custom in that

PAGE FROM JOHANN WIDMAN'S BOOK

This shows two multiplication tables from Johann's arithmetic

country to teach arithmetic in all schools. Those who were going into business went to an arithmetic school and learned from an arithmetic teacher. In this school Johann learned how to multiply in the way that Cuthbert did, although he also learned how to multiply with counters as the German merchants did at that time.

When Johann became a man he wrote an arithmetic, and in this he used the signs for plus (+) and minus (−). This was the first time these signs were ever printed.

The way in which Cuthbert and Johann multiplied was not the only one followed in those days. Some teachers ruled the paper in squares like a chessboard, writing a figure in each square, and they called their plan the chessboard method. Others ruled the paper so that it looked like an iron grating, and they called this the grating method.

These are some of the ways in which the world tried its hand at multiplying, and that is all the story we shall have to-night about Cuthbert and Leonardo and Johann.

"I don't see," said the Tease, "how they ruled the paper in the grating plan. I don't see what it means."

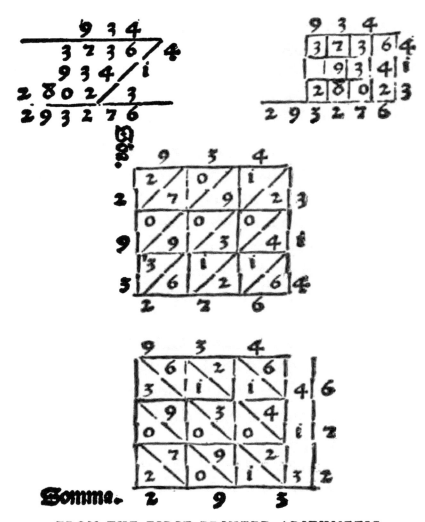

FROM THE FIRST PRINTED ARITHMETIC

This book was printed in Treviso, Italy, in 1478, a short time before America was discovered. This page shows four kinds of multiplication

"I will show you some pictures of it," answered the Story-Teller. "If you had

70

studied it in school as Cuthbert did, you would see that it is about as easy as ours."

"How is the work done?" asked Maude.

"That," said the Story-Teller, "is for you to work out some rainy day when you have

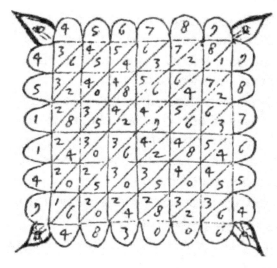

MULTIPLICATION FIVE HUNDRED YEARS AGO

From a manuscript of the fifteenth century

nothing else to do. What is much better than the grating plan at this time of the night is going to bed."

"And what do we have to-morrow night?" asked Charles.

"Think of the Question Box," mysteriously answered the Story-Teller.

THE QUESTION BOX

1. What aids in multiplying were used in ancient times by people who had numeral systems that were something like the Romans?

2. Can you tell why the ancients had less use for multiplication than we have?

3. About how long ago did Leonardo of Pisa live, and what was there peculiar about his arrangement of the work in multiplication?

4. About when was the first printed arithmetic published, and in what country?

5. Show on paper one of the ways in which they multiplied 934 by 314 when the first arithmetic was printed.

6. In what respect did Cuthbert Tonstall's work in multiplication differ from ours?

7. How did Johann Widman arrange the multiplication tables in his book?

8. Why was one kind of multiplication called the chessboard method?

9. Why was one kind of multiplication called the grating method?

10. How do you think large corporations that have a great deal of multiplying to do proceed to find products of numbers?

11. What short methods in multiplication have you learned in school?

MICHAEL STIFEL
AND HIS TEACHER

CHAPTER VII

HOW FILIPPO AND ADRIAEN AND MICHAEL DIVIDED NUMBERS

" Since this is the last story," began the Story-Teller —

" But it is n't the last," broke in the Crowd.

" Positively the last," said the Story-Teller with great emphasis, " unless — "

" Unless what ? " asked two or three voices.

" Well, you know there 's the Question Box," replied the Story-Teller, " and I don't believe one of you has answered all the questions."

" But we *can* answer them, you know, and we have three or four nights before we go home," said the Tease.

" Three or four ! Do you think I am an encyclopedia ? At any rate, let us finish up our addition, subtraction, multiplication, and division to-night. Listen, then, to the story of the way in which Filippo (fē lēp'pō) and Adriaen (äd'rē än) and Michael (mī'kå ĕl) divided numbers."

About the time that Christopher Columbus was born, a boy by the name of Filippo

HOW FILIPPO DIVIDED MDCCXXVIII BY CXLIV

He first wrote the numbers in modern figures. The rest of the work is quite different from ours

Calandri (kä län'drē) was going to school in a small Italian town. His teacher one day asked Filippo to write two numbers on a small board that he used for his arithmetic work, and he told him to write them in Roman numerals. The teacher then told him to divide the first of the numbers by the second, and of course Filippo could not do it; but he said, " I can do it if you will let me write them in our common figures."

If you had been there to watch Filippo, you would have been just as much puzzled as you would be if you saw someone dividing by the aid of counters, for the work looked very different from that used in school to-day.

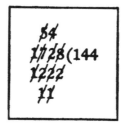

HOW FILIPPO PROVED HIS WORK

He proved that his work was right by dividing 1728 by 12 and finding 144 as the result

Filippo crossed out each figure when he was through with it, and so the work looks very strange to us.

We must not think that Filippo's way was altogether bad. He used fewer figures than we do, and if you should learn his method you might like it just as well as the one you learned in school.

After Filippo had shown that he knew how to divide, his teacher said, " Now show me how you would divide 1728 by 144, using the new method," and Filippo then divided almost the same way that we do.

Filippo lived just as Europe was beginning to pass from the old "scratch method," in which the figures were scratched out, or canceled, after they had been used, and was

```
144)1728(12
144
 28
288
288
```

FILIPPO'S NEW METHOD OF DIVIDING 1728 BY 144

Notice that Filippo used one more number (28) than we should and that he did not write the quotient above the dividend as we do. This was the second important step towards our modern method of long division

about to adopt a method very much like ours. The "scratch method" was called in Italy the "galley method" because the figures were arranged to look like a galley — a kind of boat — with its sail set. Schoolboys often showed the galley in the drawings they made in their notebooks, just as you made drawings in your books when you first went to school.

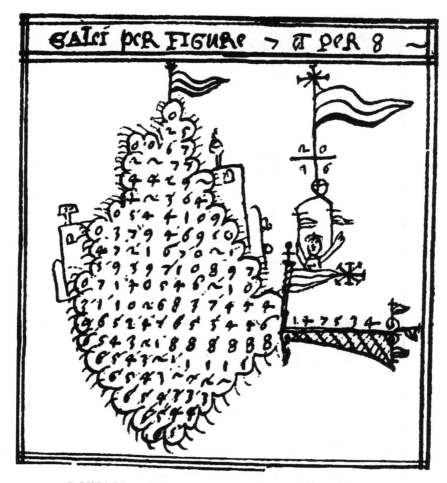

DIVISION THREE HUNDRED YEARS AGO

Page from a schoolboy's notebook written about three hundred years ago. The picture is supposed to resemble an old-fashioned galley with sails set

In using the method which his teacher called the new one, Filippo saw that it was unnecessary to write as many figures as the

77

Parti 5349 > per 83

Vienne   5349 > —— 83
          00644 - 45/83

            534
            498 |83
            ———
             369
             332
            ————
              3 >>
              332
             ————
               45
             0 45/83

Parti 3/8 p 60   Parti 13 > 1/2 p 12

3/8 — 60   13 > 1/2 — 12

0 3/8 / 0/60   13 > 1/2 / 1/12

0 3/480   Vienne 11 11/24

uienne 1/160

Parti 60 p 3/8   Parti 3/7 p 2/3

60 — 3/8   3/7 — 2/3

480 |13   3 1/2 / 2/3 |1 >

uienne 160   Vienne 0 14/28

PAGE FROM FILIPPO'S BOOK, 1491

This division of 53,497 by 83 is the first example in long division
by the modern method that ever was printed

teacher did, and so he decided that some day, when he grew up, he would write an arithmetic of his own and make the work in division still more simple.

When Filippo became a man he did what he had planned; he wrote an arithmetic containing our plan of dividing, and this was published in Florence just a year before Columbus discovered America.

In the time of Filippo Calandri, about the year 1500, Italy was much in advance of other

HOW ADRIAEN DIVIDED

The division of 150 by 6, from a work by Adriaen van Roomen, printed about three hundred years ago

European countries, and so we may expect that Filippo's improvement in division was not at once adopted outside of his country. This is the reason why a boy named Adriaen van Roomen (văn rō'mĕn), who lived in Holland one or two hundred years later, was still dividing by the old method.

For the same reason that Adriaen, going to school in Holland, was dividing by the old galley method, Michael Stifel (stē'fĕl), going to school in a little German town about the same time, also learned this same old-fashioned way, and so we see that Filippo's new method, the one that we use to-day, was not known to boys living in Holland and Germany one or two centuries later.

HOW MICHAEL DIVIDED

The division of 9552 by 12, from a work by Michael Stifel, printed about four hundred years ago

If you had lived when the Pilgrims came to America, you might all have learned division by the scratch method, because many people still used it then. Only a few years ago the Story-Teller found it still in use in Morocco, and the people who used it thought that it was better than ours. What do you think about it?

80

"I don't see why they ever used the 'scratch method,'" said Fanny. "Why did n't they use our way at first?"

"Why did n't they use electric lights instead of candles?" asked the Story-Teller, "and why did n't they have moving pictures and automobiles and airplanes?"

"Because they did not know about such things," answered Fanny.

"That is just the reason. They did not know about electric lights and railways, and neither did they know our way of dividing. As to dividing by machinery, they had never even dreamed of such a thing. Now this is enough for to-night, so get off to bed."

"But you have not mentioned the Question Box," said the Tease.

"To bed," growled the Story-Teller.

"You have n't told us how they divided with the Roman numerals," said the Tease.

"How do you think they did?" half growled the Story-Teller, trying to keep from smiling.

"Counters," said Charles.

"Right," said the Story-Teller, "and it was so hard that few could do it. And now —"

"To bed!" laughed the Crowd.

THE QUESTION BOX

1. What did Titus and Caius use to help them divide one number by another?

2. Why was the first method of dividing used by Filippo called the scratch, or galley, method?

3. How did Filippo's second method of dividing differ from the one that we use?

4. How did Filippo prove that his work in division was correct?

5. In whose book is to be found the first printed example in long division, substantially by our method?

6. Count the number of figures used by Adriaen and Michael in the galley method of dividing 1728 by 12, and

```
  $4
1728(144        144
1222        12)1728
  11         12
             52
             48
             48
             48
```

then count the number used by us in our method. Then compare the space required in the two cases. Can you see why the world refused so long to use our method?

AHMES STUDYING FRACTIONS

CHAPTER VIII

AHMES AND HERON AND JAKOB DESPAIR OF EVER LEARNING FRACTIONS

" Now we are getting near the end of our number stories," said the Story-Teller as the Crowd rushed into the room.

" Which end ? " asked the Tease. " A line has two ends."

" Never mind which is the end of a line. One thing is sure, that I am near the end of my stories."

" Yes, but *which* end of the stories ? " persisted the Tease.

" The end that we are near," said the Story-Teller, " and to-morrow night we shall be a great deal nearer."

" Then there is to be another story to-morrow night ? " said the Tease.

" That depends," replied he of the curious book. " Remember the Question Box. What shall it be to-night ? "

" I hope that it will be something that we all like," answered Maude.

" Then it must be fractions," guessed the Story-Teller, with a smile.

" Fractions ! " gasped the Crowd.

" Fractions," smiled the Story-Teller. " Just listen to the story of Ahmes and Heron (hē'rŏn) and Jakob (yä'kŏp), and see if there is not something interesting about fractions in spite of the fact that you once found them hard."

The Crowd looked doubtful, but the curious book was closed, a new log was put on the fire, and the Story-Teller began:

When Ahmes had learned from the priest in the temple by the Nile how to read and write numbers, he felt that he knew a great deal about arithmetic ; and when the priest had taught him to add and subtract, he felt that there was not much more to learn. As he grew older, however, he found that he needed to know about fractions. But no one in Egypt at that time used any fractions with a

84

numerator greater than 1, with the single exception of $\frac{2}{3}$. Instead of thinking of $\frac{3}{4}$ as we do, the priest and Ahmes thought of $\frac{1}{2}+\frac{1}{4}$; and instead of thinking of $\frac{7}{8}$, they thought of $\frac{1}{2}+\frac{1}{4}+\frac{1}{8}$.

So it is no wonder that Ahmes had a great deal of trouble in learning fractions, and it is no wonder that he never learned how to work with fractions as we work with them.

Not only did the people of that time use only fractions with numerators 1, like $\frac{1}{2}$, $\frac{1}{3}$, $\frac{1}{4}$, and $\frac{1}{8}$, but for more than two thousand years after Ahmes died these fractions were commonly used in Egypt. They were also used in Babylon and in various other countries.

Nearly two thousand years after Ahmes studied in the temple on the banks of the great river which makes Egypt the fertile country that it is, there lived in Alexandria, at the mouth of the Nile, a boy named Heron. He was interested in machines and in measuring heights and distances, and he made friends with the scholars in Alexandria who studied the stars. He visited the Great Pyramid, listened to the stories of its building and its purpose, and came in contact with many travelers at this great port on the Mediterranean. So Heron grew up in very interesting surroundings and at an interesting time in the world's history —about the beginning of the Christian era.

When Heron went to school he found that he needed another kind of fraction than the one used by Ahmes. When he came to the very careful measurements used in making machines or in finding the position of the stars, he found that he needed certain other fractions that had come into use in Alexandria long after the time of Ahmes. These fractions had 60, or 60 × 60, or 60 × 60 × 60 for their denominators, and since everyone knew this, it was not necessary to write the denominators. When we write 0.5 we mean $\frac{5}{10}$, and it is not necessary to write the denominator. To use our modern terms we may say that 23 minutes meant $\frac{23}{60}$ of something that was being measured, and 23 seconds meant $\frac{23}{60}$ of a minute, or $\frac{23}{3600}$ of the thing that was being measured. In the same way the next 23 meant $\frac{23}{60}$ of a second, and so on. In this way it was possible to have fractions without writing the denominators at all.

Do you think that this is about the most impractical thing you ever heard of? If you do, just remember that you use these fractions whenever you speak of the time of day. If it is 2 hr. 25 min. 47 sec. past noon, it is really 2 hr. $+\frac{25}{60}$ hr. $+\frac{47}{3600}$ hr. past noon. In other words, you are using the awkward fractions found in the writings of all who needed to use arithmetic for very accurate computation in the time of Heron of Alexandria, one of the great writers of Egypt at the beginning of the Christian era.

There have been many other ways of writing fractions, and I will tell you of one used by a boy named Jakob who lived in Germany about four hundred years ago.

Jakob went to school to an arithmetic teacher and tried his best to learn the new Arab way of writing fractions. The Arabs seemed to have learned these fractions from the Hindus. So Jakob learned how to write $\frac{3}{4}$ and $\frac{7}{8}$ and thought he knew a great deal about fractions.

Jakob's teacher did not think very highly of what he called " the new-fangled numerals," meaning the common ones that we use. He thought that the old Roman numerals, which were commonly used in Germany at that time, would continue to be used.

So Jakob learned how to write the Arabic fraction, but he then did a very funny thing —

A PAGE FROM JAKOB KÖBEL'S BOOK

This shows Jakob's strange way of writing fractions

he made up his mind that he would write an Arabic fraction with Roman numerals, and you can imagine how strange it must have looked.

No Roman ever wrote a fraction like that, and no one ever put such an awkward fraction

88

in a book until Jakob wrote an arithmetic himself. He thought he had done something very great, but no one else thought so.

It was nearly a hundred years after Jakob's time that a boy named Simon went to school in the city of Bruges (broo'jĕz), in Belgium. Like Ahmes and Heron and Jakob, he wrestled with fractions; but while these three boys despaired of ever learning how to conquer them, Simon made up his mind that he could come out the victor, and he did.

Simon saw that the Arabic fractions, which had then come to be called common fractions, were all right for cases like $\frac{3}{4}$, $\frac{7}{8}$, and $\frac{5}{12}$, but that they were not good for much when very fine measurements were required. So he set about to get a better kind of fraction for such work, and when he became a man he wrote a book on decimal fractions, printed more than three hundred years ago. You will hardly think that he knew much about decimals when you look at a page from his book, but he was the first man to write a work upon the subject.

It was not long before Simon's ideas were improved, so that the world soon came to write decimal fractions about as we do to-day. We have not yet agreed upon the decimal point, however, for children who go to school in England write decimal fractions with the point halfway up (3·14), and in the rest of Europe they use a comma for a decimal point (3,14).

Explication du donné. Il y a trois ordres de nombres de Difme, defquels le premier 27 ⓪ 8 ① 4 ② 7 ③, le deuxiefme 37 ⓪ 8 ① 7 ② 5 ③, le troifiefme 875 ⓪ 7 ① 8 ② 2 ③.

Explication du requis. Il nous faut trouver leur fomme. *Conftruction.* On mettra les nombres donnez en ordre comme ci joignant, les aiouftant felon la vulgaire maniere d'aioufter nombres entiers, en cefte forte:

⓪	①	②	③
2	7	8	4	7
3	7	6	7	5
8	7	5	7	8	2
9	4	1	3	0	4

HOW SIMON STEVIN WROTE DECIMALS

This is the first book ever published on decimal fractions. The first number is 27.847

So the world's ideas grow just as yours grow, and fashions change as the years go on. For this reason the fashion of decimal points varies from time to time and from country to country very much like the fashions of men's collars, of women's hats, of cutting boys' hair, and the ways of wearing belts on a girl's dress.

90

"Do you mean that we have fashions in arithmetic just as we have fashions in clothes?" asked Emily.

"Certainly; why not? We have been seeing fashions in reading and writing numbers, in adding, subtracting, multiplying, and dividing. Why do you call the answer in multiplication a product and not a sum? Nothing but fashion! It was once the fashion to use the word 'product' for the answer in addition, and the word 'sum' for the answer in subtraction. It is the fashion in school to speak of a dividend as a number to be divided, but it is not the fashion with business men to use it in this way. It is the fashion for you all to go to bed at this time, and —"

"But you say that fashions change," said the Tease.

"This fashion does n't change at all," replied the Story-Teller.

"And the fashion of telling us number stories at night does n't change, either," said the Tease.

"Neither does the fashion of the Question Box," laughed the Story-Teller as the Crowd started for bed.

THE QUESTION BOX

1. The kind of fraction used by Ahmes is called the unit fraction. Why does it have this name?

2. Ahmes used a fraction which we might write with our symbols $\frac{1}{2} + \frac{1}{4} + \frac{1}{8}$. What single fraction is this?

3. What single fraction is $\frac{1}{2} + \frac{1}{3}$?

4. About when did Heron live? What new kind of fraction did he use?

5. When we write 4 hr. 10 min. 30 sec. we mean 4 hr. $+ \frac{10}{60}$ hr. $+ \frac{30}{3600}$ hr. How do you express this as 4 hr. plus a single common fraction in lowest terms?

6. In hours, minutes, and seconds given in Question 5 how do you express the result as a decimal fraction to the nearest thousandth of an hour?

7. In what curious way did you find that Jakob wrote common fractions?

8. Who wrote the first book on decimal fractions? This was about how long ago?

9. In the first printed book on decimal fractions how was 27.847 represented?

10. In what part of the world do they write 3.14 for $3\frac{14}{100}$? Where is this written 3·14? Where is it written 3,14?

11. Which fractions do you find the easier to work with, decimals or common fractions? Think of the case of $\frac{29}{125}$ and 0.232; then of the case of $\frac{1}{8}$ and 0.125.

92

CHAPTER IX

NUMBER PUZZLES BEFORE THE LOG FIRE[1]

"So this is to be our last night," began the Story-Teller.

"The last but one," said the Tease.

"Are you sure?" asked the Story-Teller.

"Sure!" answered the Crowd, not intending to miss what would come by the asking.

"Well, what shall it be to-night? How would you like a nice long story about square root or cube root?"

"Cube root does n't sound very interesting," said Charles.

"Logarithms?" asked the Story-Teller.

"We never heard of them," said Maude. "Tell us something interesting."

[1] Chapters IX and X are not intended to be read aloud.

"Well, since we have a good log fire here, perhaps you would all like to hear the story of 'Number Puzzles before the Log Fire.' Listen, then, and look for Ching and Chang and An-am and all our friends, for they are all coming to visit us to-night.

"A curious thing happened one cold night, the place being the edge of a dense forest not many miles from here, the house being built of logs, the room being low and having a huge fireplace, the furniture consisting of a number of comfortable chairs arranged about the fire, and the time being early candlelight.

"It may well be thought that there was developed in the forest a certain amount of curiosity in the fact that a fire burned brightly in a house which contained no living soul and that plenty of candles lighted up a room for no apparent purpose; for the animal life of the forest has quite as much curiosity as you and I and most other people have, and this is the reason that into the window there looked the squirrels, the owls, and the rabbits that made their homes in the trees or in the burrows of the woods.

94

"Over the heavy door of the log house someone had written in the years gone by these words: 'Let no one without imagination enter here.' And so the door is closed to you and me and all the world unless we have imagination. For us the log fire is cold and the candles do not burn and the heavy door is barred if we belong to that humdrum class that never has any daydreams. You, then, who are without imagination and who belong with those unfortunates who believe only what they see, stop on the threshold. For you the log fire has no attraction and the empty chairs will have no occupants."

From out of the Great Unknown of Time and Space there come, one by one, into the room of the house of logs, each taking a chair in front of the great fire, our friends of long ago: Ching and Chang and Wu from the Land of the Yellow Dragon; An-am and Lugal from the Tigris and Euphrates; Menes and Ahmes from the banks of the Nile; Hippias from Greece; Titus and Caius from the Seven Hills of Rome; Daniel from the Mount of Olives; Gupta, Mohammed, Gerbert, and most of the others of whom we have read in our stories of Number Land, — strangers to each other, strangers to our present civilization, but drawn together by their interest in the puzzles of the world of number.

For this is to be a night of number puzzles, and each is to bring his contribution, each is to add to the interest of boys and girls in some of the curious things that have grown up in the thousands of years during which the world has played as well as worked in the realm of arithmetic. Imagine, therefore, Ching in his dress of leopard skin, and An-am with his cloak of sheepskin, and Ahmes with his roughly woven linen robe, and Gupta with his brown skin and his brown garb, and Mohammed with his long white robe, and all the rest, each in the dress of his land and time, all seated in front of the blazing logs, all intent upon hearing some of the number puzzles of the ages.

"In my land," said Ching, "there is an ancient book, perhaps the oldest book in all the world, and the name of this is Yih King (yē kĭng). It was written thousands of years ago and records one of the earliest of all the

curious things connected with number, for it
says that once upon a time there came out of

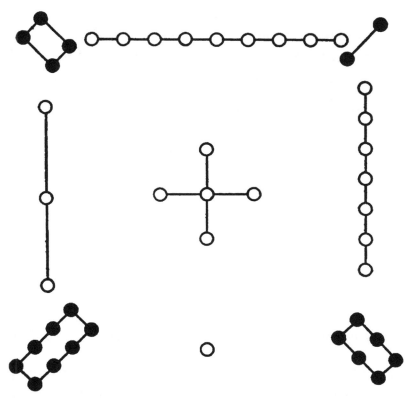

THE WORLD'S OLDEST MAGIC SQUARE

From the Yih King, one of the oldest Chinese books, written
perhaps three thousand years ago

the Yellow River a large turtle, and on its
back were some strange marks which puzzled
everyone who saw them. These I have written
on paper for you to see."

97

Then Ching passed around, that all might see it, a paper with some very curious figures written upon it.

"I know what it is," said Gupta; "it is a magic square, and the dots are numbers. The columns add to 15 and so do the rows and so do the diagonals. It is the world's oldest

4	9	2
3	5	7
8	1	6

A MAGIC SQUARE

The numbers of the dots in the magic square in the Yih King

number mystery. It is used as a charm all through the East; and in the Middle Ages, long after I died, it was used in many parts of Europe to drive away disease and to bring good fortune."

Gupta was right, for in the magic square is one of the oldest and most interesting curiosities of number to be found anywhere in the world.

98

"When I was living in the early days of Egypt," said Menes, "we did not have many number puzzles, but since then I have watched the world and have seen some strange things happen. One of the most curious relates to the numbers that are now used in Europe and America and in all places under their influence. If you take any number, say 3476, and reverse it, 6743, and then take the difference, 6743 − 3476 = 3267, this difference will always be exactly divisible by 9. Whatever number you take, this will always be true. You may think that you can find a number with which this is not true, like 2222, but you will see that the difference will always be divisible by 9."

```
6743
3476
────
3267
```

All the company tried to find out if Menes was right, using different numbers, but in every case the result was exactly divisible by 9. Chang, however, made a mistake. Having started with 3827, he simply mixed up the figures and obtained 2783, but he found that the difference was still divisible by 9.

```
3827
2783
────
1044
```

99

"Yes," said Menes, "you may mix up the figures in any way and the rule still holds."

Some of the company knew the reason; some of you may know it; all of you will easily see it after you have studied a little algebra, for algebra is like a great electric light — it reveals many of the strange secrets of arithmetic.

"One of the most curious things that I have noticed in looking at the world for the past two thousand years," said Heron, "is seen in a simple example in addition. If you write any three numbers, say numbers of four figures each, I will at once write three numbers underneath them and will tell you before I do so the sum of all six numbers. It does not make any difference what numbers you take, nor does it make any difference whether they are all different or all alike."

This struck the company as a very curious thing, especially as Heron added, "Whatever three numbers you write, the sum of the six numbers will be 29,997."

```
2376
4152
3804
```

Adriaen then wrote on a piece of paper the three numbers 2376, 4152, and 3804.

Underneath these numbers Heron at once wrote the numbers 7623, 5847, and 6195.

He then told the company to add the six numbers, and, to the surprise of all, the sum was what Heron had prophesied.

Cuthbert Tonstall said he knew the trick. Do you know it?

If Adriaen had written four numbers, Heron would have written four, but he could have done the trick just the same, although he would have stated another sum.

2376
4152
3804
7623
5847
6195

" A strange thing was shown to me when I went as a boy to Barcelona," said Gerbert, "more than nine hundred years ago, and I will show it to you because it may prove to be interesting. If you will write on paper any number of as many figures as you please, I will write a single figure at the end of your number, and the number will then be exactly divisible by 11."

This did not seem a possibility, and so Jakob wrote 74,289.

Gerbert simply glanced at the number, wrote 6 after it, and said, " You will now find that 742,896 is exactly divisible by 11, and I have given you good measure, for I have also made it divisible by 22, 33, and 99."

The company all tried it and found that Gerbert was right.

Titus then wrote 66,742, and Gerbert wrote 5 at the end, saying, "You will find that 667,425 is exactly divisible by 11, and this time I have made it also divisible by 33 and 55."

The question now is, How did Gerbert do the trick?

"I think," said Hippias, "that you may be interested in a curious thing that I noticed a few hundred years ago. Two boys were giving each other puzzles, and one said, 'Think of any number you wish, multiply it by 2, add 18, divide the result by 2, and subtract the original number of which you thought.'

'I have done all this,' said the other.

'Then the result is 9,' said the first boy.

"Now," continued Hippias, "I suggest that all the company try this trick."

Then each one took a number, and each

did as directed, and to the surprise of all, the result in every case was 9.

"I know the trick," said Jakob. Of course he did, for he knew algebra. "I will give you a still better one," he continued. "Take any number, multiply it by 6, add 12, divide by 3, subtract 2, divide by 2, subtract the original number, and add 9."

"Well," said Adriaen, "what of it?"

"The result is 10," said Jakob.

At this the whole company laughed, for each had started with a different number, and yet each had ended with 10.

"I remember," said Leonardo, "that an old schoolmaster used to ask the new pupils which was correct, '6 and 7 *are* 14' or '6 and 7 *is* 14'"; but the rest of the boys were too quick to be puzzled by anything like this.

"Did you ever run across this ridiculous problem?" said Filippo. "If 6 cats eat 6 rats in 6 minutes, how many cats will it take to eat 100 rats in 100 minutes, at the same rate? It used to puzzle us down in Florence four or five hundred years ago."

"I know," said Wu; "it will take 100 cats." Was he right?

"When I was a boy," said Michael Stifel, "we had a couplet which ran like this:

Ten fingers have I on each hand
Five and twenty on hands and feet.

You know we used to say 'five and twenty' for twenty-five. The statement is perfectly true, but how do you explain it?"

At this the company lapsed into silence for a time.

"But I don't have 'five and twenty' fingers on my hands and feet," insisted Wu.

"You must think of your toes as fingers," said Robert.

"It is easy enough if you only know how," said Michael.

Then Heron said, "The explanation is —"

"Don't give it!" called out several voices. "We want to think it out for ourselves."

And so I shall have to leave it for you.

"Can any of you write an even number, using only odd digits?" asked Adriaen.

"I can," said Cuthbert. "It is just as easy as writing a hundred without using zero."

"Well, don't tell us," said Gupta. "We also want to think this out for ourselves."

"I believe," said Jakob, "that one of the oddest things I ever saw in numbers is a set of products that I have written here on paper"; and saying this he passed the paper around the room so that all might see it:

$$
\begin{array}{lll}
3 \times 37 = 111, & \text{and} & 1+1+1=3 \\
6 \times 37 = 222, & \text{and} & 2+2+2=6 \\
9 \times 37 = 333, & \text{and} & 3+3+3=9 \\
12 \times 37 = 444, & \text{and} & 4+4+4=12 \\
15 \times 37 = 555, & \text{and} & 5+5+5=15
\end{array}
$$

"Now," he said, "how many can tell me the product of 18 and 37 and what the sum of the three digits is in the answer? Also, what are the products of 21 and 37, of 24 and 37, and of 27 and 37, and what are the sums of the digits in the various answers?"

Can you answer these questions?

Wu now wrote the following on paper:

$$
\begin{array}{l}
7 \times 15{,}873 = 111{,}111 \\
14 \times 15{,}873 = 222{,}222 \\
21 \times 15{,}873 = 333{,}333
\end{array}
$$

He then passed the paper around the room so that all the boys might look at it.

"And now I ask," he said, "the results of 28 × 15,873, of 35 × 15,873, of 42 × 15,873, and of 49 × 15,873."

Several could tell the results. Can you tell what they are?

Wu let them think for a few minutes and then asked: "Can any of you tell by what I should multiply 15,873 to have 888,888? to have 999,999?"

About half of the company could give the answers. Can you?

"While you are talking about such curious things," said An-am, "I will show you something interesting."

$$1 \times 8 + 1 = 9$$
$$12 \times 8 + 2 = 98$$
$$123 \times 8 + 3 = 987$$
$$1234 \times 8 + 4 = 9876$$
$$12345 \times 8 + 5 = 98765$$
$$123456 \times 8 + 6 = 987654$$
$$1234567 \times 8 + 7 = 9876543$$
$$12345678 \times 8 + 8 = 98765432$$
$$123456789 \times 8 + 9 = 987654321$$

He then wrote the above figures on a piece of paper and passed it around the room.

"Yes," said Menes, "that is interesting, but I will show you something that seems to me even more curious."

He then took a piece of paper and wrote:

$$1 \times 9 + 2 = 11$$
$$12 \times 9 + 3 = 111$$
$$123 \times 9 + 4 = 1111$$
$$1234 \times 9 + 5 = 11111$$
$$12345 \times 9 + 6 = 111111$$
$$123456 \times 9 + 7 = 1111111$$
$$1234567 \times 9 + 8 = 11111111$$
$$12345678 \times 9 + 9 = 111111111$$

"It seems to me," said Ahmes, "that these figures which were invented many hundred years after I died are very funny things. You could never do that with the numerals that I learned in Egypt four thousand years ago."

"No," answered Cuthbert, "and everyone knows that you couldn't do anything else with them, they were so bad!"

"Well," replied Ahmes, "you need not be so proud of your numerals; try writing twelve thousand twelve hundred and twelve."

This interested the company, and you will find it worth trying.

"Speaking of twelves," said Leonardo, "which is greater, six dozen dozen or a half dozen dozen?" Which is it?

"When I was a boy," said Titus, "I was given a puzzle problem which I will change a little so as to give it in English. Take nine from six, ten from nine, and fifty from forty, and show that the remainder is six."

This puzzled the company for a few minutes. Then Caius said, "It is very easy; all you have to do is to —"

"Don't tell us," cried the company.

"I have something that is as curious as those multiplications of An-am and Menes," said Lugal. "Look at this," and he wrote the following numbers and signs on paper:

$$9 \times 9 + 7 = 88$$
$$98 \times 9 + 6 = 888$$
$$987 \times 9 + 5 = 8888$$
$$9876 \times 9 + 4 = 88888$$

This struck all the company as very strange, because the left-hand numbers were 9, 98, 987, and 9876, and the numbers added were 7, 6, 5, and 4, while the products were always made up of 8's.

"Can you tell me," Lugal added, "the values of each of the sets of numbers which I will now write on paper?"

Saying this he wrote the following numbers:

$$98{,}765 \times 9 + 3 \qquad 9{,}876{,}543 \times 9 + 1$$
$$987{,}654 \times 9 + 2 \qquad 98{,}765{,}432 \times 9 + 0$$

Several of the company could tell the results at once. Can you do so?

"I ran across a curious thing about the number 45 when I was living in Spain a few hundred years ago," said Gerbert.

"What is it?" asked several.

"Well, 45 is equal to $8 + 12 + 5 + 20$, and these four numbers, 8, 12, 5, and 20 have curious combinations with 2, thus:

$$8 + 2 = 10, \quad 12 - 2 = 10, \quad 2 \times 5 = 10, \quad 20 \div 2 = 10,$$

the result of each of these four different operations being 10. Did you ever know anything more curious?"

"That. is certainly very odd," replied Leonardo; "but I know another strange thing about 45. If you take the number 987,654,321, made up of the nine digits, reverse it, and subtract like this, you will have three numbers, — the minuend, the subtrahend, and the remainder, — and the sum of the digits of each of the three is exactly 45."

$$\begin{array}{r} 987,654,321 \\ 123,456,789 \\ \hline 864,197,532 \end{array}$$

The whole company thought that 45 was certainly a very curious kind of number.

"Here is a good one," said Adriaen. "Show how to write one hundred, using only the nine digits and the signs of arithmetic."

This puzzled everyone, and so Adriaen showed the company the following:

$$100 = 1 + 2 + 3 + 4 + 5 + 6 + 7 + 8 \times 9$$

"Easy enough, is n't it?" said Adriaen. "I know another way, too. Just take —"

"Don't tell us! Give us time to think about it," cried out several.

"While some of you are thinking about Adriaen's problem," said Johann, "perhaps others who are here would like to find a number of two digits which is equal to twice the product of those digits."

"I know," said Ching, "it is 12."

"No," said Johann, "for 12 is not equal to twice the product of 1 and 2."

"Well, I know," said Chang. "It is —"

"Don't tell us," said several who were figuring it out on paper.

Wu found the answer first. What is it?

"Here is a nice little problem for you," said Michael: "A snail crawling up a pole 10 feet high climbs 3 feet every day and slips back 2 feet every night. How long will it take to reach the top?"

"Ten days," said Wu — but he was wrong.

"Here is a little trick," said Jakob:

```
    I C C          I N U
      I N            N U
    -----          -----
    N T T          L N U
    I C C          N U S
    -----          -----
    I A N T        O I N U
```

"Write figures in place of these letters so as to make the multiplications correct. Since it will take you some time to find these figures, here are two tricks in division also:"

```
              E M A                        M T
    M A)U E M A            E M A)U U S S
        M A                        M A S
        ─────                      ─────
          T M                        O S S
          A S                        I A S
        ─────                      ─────
        E M A                        A S
        E M A
        ─────
```

"These will take a great deal of time," continued Jakob, "so you had better let them go until to-morrow. In place of them you may take this little puzzle: A watermelon weighs ⅘ of its weight and ⅘ of a pound. How many pounds does it weigh?"

"That reminds me," said Lugal, "of this problem: If a herring and a half costs a cent and a half, how much will a dozen and a half herrings cost?"

"I wonder," said Gupta, "how many of you can write nineteen, using only four 9's."

"That is too easy," said Hippias. "It is no harder than to write two with four 9's. Why don't you ask how many of us can write twenty, using only four 9's? That is a puzzle worth thinking about."

"And that reminds me of one which I heard many years ago," said Titus. " A bottle and a cork cost $1.10, and the bottle costs $1 more than the cork. How much does each cost?"

"The bottle costs $1, and the cork costs 10 cents," said Ching.

"Wrong," said Titus. "I don't think you could even tell which is the heavier, a pound of gold or a pound of feathers."

"They weigh just the same," said Ching.

"I told you that you couldn't tell," said Titus. "I don't believe you could even tell how many minutes it would take to cut a strip of cloth 50 yards long into strips 1 yard long if each cut takes one minute."

"It would take," began Ching; but then he remembered that the other time he had spoken without thinking, so he stopped.

"This problem," said Caius, "reminds me of one which I heard many years ago. After cutting off $\frac{1}{10}$ of a piece of cloth, a merchant had 100 yards left. How many yards did he have at first?"

" He had 110 yards," said Ching. But Ching spoke again before he really solved the problem.

"You had better take a little more time," remarked Caius.

While Ching was thinking, Hippias told a story about himself.

"I was caught," said he, "about two thousand years ago by a simple little question. My teacher in Athens asked me how many quarter-inch squares it would take to make an inch square. I got this all right; but when he asked me how many quarter-inch cubes it would take to make an inch cube, I could n't tell. Can you tell?"

"While I am thinking," said Daniel, "you may care for this little problem: If an apple just balances in weight with $\frac{3}{4}$ of an apple of the same weight and $\frac{3}{4}$ of an ounce, how much does the apple weigh?"

"Speaking of apples," said Ahmes, "two fathers and two sons divided three apples among themselves, each receiving exactly one apple. How was this possible?"

" It is n't," said several.

"Oh yes, it is," said Ahmes, "if you — "

" Don't tell us," cried the whole company.

114

"It reminds me," said Mohammed, "of a puzzle I heard about a thousand years after I lived in Bagdad. In a family party there were 1 grandfather, 2 fathers, 1 grandmother, 2 mothers, 4 children, 3 grandchildren, 1 brother, 2 sisters, 2 sons, 2 daughters, 2 married men, 2 married women, 1 father-in-law, 1 mother-in-law, and 1 daughter-in-law. How many were there in the party?

"Let me see," said Wu; "$1 + 2 + 1 + 2 + 4 + 3 + 1 + 2 + 2 + 2 + 2 + 2 + 1 + 1 + 1$ is 27."

"You have come nowhere near it," replied Mohammed.

"Then how many do you say there were?" asked Wu.

"I know how many there were," said Daniel, "You see it was like this: there were —"

But just then a deep-toned bell out in the forest tolled twelve; and when the owl looked in the window the fire was out and every chair was vacant and not a soul was in the room, and the only noise was that of the winking of the owl's eyes and the breathing of the squirrel and the soft step of the rabbit as he found his way to the hole under the roots of the oak tree near the log cabin.

THE QUESTION BOX

1. What is a magic square? Where is it first found? How has it been used?

2. Can you arrange a magic square of nine figures in a different way from the one shown by Ching and Gupta?

3. If you take a number of two figures and reverse it, is the difference between the numbers always divisible by nine? Give five cases in which this is true.

4. Can you tell how to find out, without dividing, whether or not a number is exactly divisible by nine?

5. Find a number which when multiplied by 3 and the result added to 8 becomes 35.

6. Explain this old puzzle:

> Every lady in this land
> Has twenty nails upon each hand
> Five and twenty on hands and feet
> And this is true without deceit.

7. Write 78, using only the figure 7, but repeating it as often as you wish.

8. When the bell in the forest struck midnight, having struck the time every hour of the twenty-four, how many strokes had it made since the preceding midnight?

9. If there are 2 fathers and 2 sons in a room, and no one else is there, what is the smallest number of persons that can be in the room?

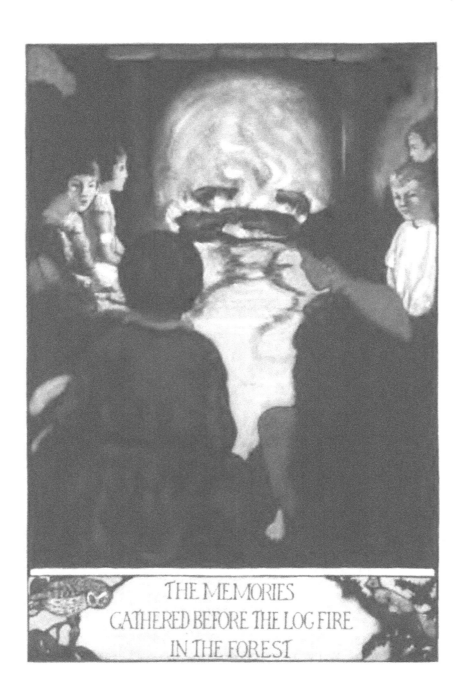

THE MEMORIES
GATHERED BEFORE THE LOG FIRE
IN THE FOREST

CHAPTER X

CURIOUS PROBLEMS BEFORE THE LOG FIRE

"Have you imagination still?" asked the Story-Teller, "or are you too proud to confess that you have any? Are you among those unfortunates who were barred from the log cabin when last there gathered the group of boys from many lands and many times? For if you have no imagination, or if you are afraid to confess that you have any, you should turn your back, for here again is the log cabin by the edge of the forest.

"The owl has been there, and the fire burns. The squirrel has looked in the window and seen the empty chairs. The rabbit has listened in vain for any footfalls. And yet, somehow, in some mysterious way, out from

the Somewhere have come our friends of old
— Ching and Chang and An-am and Lugal
and all the rest. Where a moment ago were
empty chairs, there are now chairs filled with
the boys of thousands of years ago, of yester-
day, and of to-day, each with a pencil and a
pad of paper, and each with some curious
problem of the long ago. I suppose the
boys themselves were not there — just their
Memories; but this is the realm of Imagination, so let us not be like those stupid ones
who will not believe that Memories can talk."

" The problem that I have brought," said
Mohammed, " is a very old one that I learned
from an Arab more than a thousand years
ago. It is so easy that it is hard; but if you
will not try to make it hard, it is very easy.
It is easier to make it hard than it is to make
it easy, however, for most people find it quite
hard to make it easy."

" Well, if it is as hard to make it easy as it
is to make your statement easy, then it really
must be awfully hard," said Cuthbert.

"We don't want any easy ones," said Cuthbert, " so give us a good puzzle."

118

" There were two Arabs who sat down to eat," said Mohammed, " one with five loaves, and the other with three, all the loaves having the same value. Just as they were about to begin, a third Arab came along and proposed to eat with them, promising to pay eight cents for his part of the meal. If they ate equally and consumed all the bread, how should the eight cents be divided? Remember that one Arab furnished five loaves and the other furnished three loaves."

" While you are thinking of that," said Cuthbert, " here is another good one. A man had five pieces of chain, each having three links. He asked a blacksmith how much he would charge to make them into one piece of chain. The blacksmith replied that he charged two cents to cut a link and two cents to weld or fasten together a link. How much was the blacksmith's charge?"

The company disputed for some time, and could not agree upon the answer.

Finally Lugal said: "I have it! You see it is this way: all you need do is —".

"Don't tell," said the rest of the boys.

"I will give you a little problem," said

Johann, " but I warn you that there is a trick in it. How may you arrange the figures 1, 6, 8 so that the three-figure number that you make shall be exactly divisible by 9 ? "

" It can't be done," said Filippo, after a moment's thought.

" Yes, it can," said Johann, " but it is a trick, and you may think about it until to-morrow."

" I remember in my day," said Leonardo, " a very famous problem that I first heard from an Arab scholar. He told me that when chess was invented, the Persian king was filled with great joy and commanded that chessboards should be placed in all the temples of the kingdom.

"' Moreover,' he said to the inventor of the game, 'ask what you will.'

" Thereupon the inventor replied, ' Give me, O king, a single grain of wheat to place in the first square of the chessboard, two in the second, four in the third, and so on, doubling the number each time, until all the sixty-four squares are filled.'

" You will be surprised, I think, to know how many grains of wheat will be necessary."

120

"How many are there?" asked Ching.

"More than a hundred," said An-am.

"Rather!" said Leonardo; "for you have $1 + 2 + 4 + 8 + 16 + 32 + 64 + 128 + 256 +$, and so on, for the sixty-four squares, and since the next three numbers are 512, 1024, and 2048, you see that they are rapidly becoming very large. The total number of grains is enormous; it is 18,446,744,073,709,551,615, and this would make more wheat than there is to be found in all the world."

"I remember," said Cuthbert Tonstall, "a problem somewhat like that. According to this problem a man wanted a blacksmith to shoe his horse. The blacksmith said he would do so if he were paid one cent for the first nail, two cents for the second, four cents for the third, eight cents for the fourth, and so on, doubling the amount each time. There being twenty-four nails, how much did it cost to shoe the horse? The answer is—"

"Don't tell us," cried several of the boys. "We can work out such a simple problem as that for ourselves."

"Here is an interesting problem that I remember to have seen many years ago," said

Michael. "A girl was carrying a basket of eggs, and a man driving a horse hit the basket and broke all the eggs. Wishing to pay for the damage, he asked the girl how many eggs she had. The girl said she did not know, but she remembered that when she counted them by twos, there was one egg left over; when she counted them by threes, there was also one egg left over; when she counted them by fours, there was also one left over; but when she counted them by fives, there were no eggs left over."

"'Well,' said the man, 'I can now tell you how many you had.'

"I wonder," continued Michael, "how many of you can tell the number of eggs?"

"I think," said Adriaen, "that there is more than one answer."

"Yes," replied Michael; "but the girl probably had not more than four or five dozen eggs in the basket, and there is only one answer less than these numbers."

"We can find the answer," said several of the company.

And so I leave it for you.

"When I was a boy," said Titus, "I remember to have heard a simple little problem that was

122

brought to Rome from Alexandria, in Egypt:
A mule and a horse were carrying some
bales of cloth. The mule said to the horse,
'If you give me one of your bales, I shall
carry as many as you.'

"'If you give me one of yours,' replied the
horse, 'I shall carry twice as many as you.'

"How many bales was each carrying?"

"Yes," said Leonardo, "I heard the same
one when I was a boy, and when I became
a man I put it in my arithmetic."

"The answer," said Titus, "is —"

"Oh, that is too easy; we can work it out,"
said Cuthbert, and this met with the approval
of the company, so I leave it for you.

"Although I was born in Florence," said
Filippo, "my parents took me once to Venice.
There I found that the clocks struck from 1 to
24 instead of striking from 1 to 12. I was much
interested to know how many strokes a clock
made in twenty-four hours, and so I added
the numbers from 1 to 24. But I afterwards
learned that there was a shorter way of finding the answer. Do you know this way?"

At this the company was silent for a
minute, when Leonardo remembered that he

had it in one of the books that he wrote when he grew up to be a man, and Cuthbert remembered that it was in his book, and Michael had it in his; so perhaps it is just as well to leave it for some of you to find out.

"It is getting late," said Robert Record, "and we have n't time for any more problems."

"Just a few more," cried the company.

"Well," said Robert, "who has a few short problems that will set us thinking?"

"I have one," said Caius.

"All right; go ahead," said Robert.

"I heard this from a Greek teacher who came to Rome when I was a boy," said Caius. "I was playing in the Forum one day when this Greek teacher came along and stopped where several of us were throwing a ball from one to another. He asked me how long it would take Achilles (à kĭl'ēz) to overtake a turtle that had a mile the start. I told him it would depend on how fast Achilles ran.

"'Well,' said he, 'suppose that Achilles ran ten times as fast as the turtle and that he ran a mile in ten minutes.'

"'Then,' said I, 'it would take him about fifteen minutes.'

"'No,' said he, 'Achilles could never overtake the turtle.'

"I did not like to contradict him, but I said that I felt sure that I could catch up with a turtle in a very short time.

"'This is how it is,' he replied. 'When Achilles reaches the place where the turtle was, the turtle will be $\frac{1}{10}$ of a mile ahead, because the turtle runs $\frac{1}{10}$ as fast as Achilles.'

"'Yes,' said I.

"'And when Achilles reaches the place where the turtle then is, the turtle will be $\frac{1}{100}$ of a mile ahead.'

"'Yes,' I replied.

"'And when Achilles reaches the place where the turtle then is, the turtle will be $\frac{1}{1000}$ of a mile ahead.'

"'Yes,' said I, 'but $\frac{1}{1000}$ of a mile is only about five feet.'

"'Yes, but don't you see that when Achilles reaches that point, the turtle will be $\frac{1}{10}$ of that distance ahead?'

"'Yes,' I replied.

"'Then,' said he, 'because Achilles is continually reaching some place where the turtle was, and the turtle is then always $\frac{1}{10}$ the last distance ahead, Achilles can never catch up with the turtle.'

"It seemed too much for any of us boys, but we gave up playing ball and began to think about the Greek's problem. That was nearly two thousand years ago, and I have n't yet been able to see how Achilles could ever catch up with that turtle. It seems to me that the turtle will always be a very little ahead of Achilles."

"I don't know how to help you," said Cuthbert, "but I will give you something else to think about. Can you tell me how to distribute ten pieces of sugar in three cups so that every cup shall contain an odd number of pieces?"

"It can't be done," said Wu.

"That is because you don't know how," replied Cuthbert. "Try this: In a box there are six apples. Divide these equally among six boys in such a way as to leave one apple in the box, but do not cut any apple."

While some of the boys were thinking about Cuthbert's problem, Wu gave this one: "A man sold his farm for $5000, which is what it cost him, then bought it back for

$4500, and then sold it again for $5500. How much did he gain?"

"He gained $1500," said Adriaen.

"Wrong," replied Wu.

Can any of you tell how much he gained?

"I will give you a good puzzle," said Ching. "What is the number of two figures which multiplied by 8 becomes 20?"

"There isn't any such number," said Lugal.

"There must be some catch about it," said An-am.

"Yes, but it is a nice catch after all," answered Ching, "and I will leave it for all of you to think about."

"Did you know that six and six are eleven?" asked Titus.

"That is one of your old tricks with Roman numerals," said Cuthbert. "If you can show that, you can show that four and four are nine and that seven and seven are twelve."

"Of course I can," said Titus.

The rest of the boys thought that all this was too easy. What do you think?

"Here is a better one, then," said Titus. "Three brothers divided four apples among themselves so that one had no more than the others, and yet no apple was divided. Explain how this was possible."

" It can't be done," said Menes.

" Oh, yes, it can," said Titus.

" Well, then, there is some trick about it," said Jakob.

" Of course there is," answered Titus, " but it is a good trick and you can't see it."

" But each must have had an apple and a third," said Chang.

" No, because the apples could not have been divided equally."

" Well, I don't see it at all," said Chang, " and I want to think about it further."

The rest of the boys laughed, because no one could see how it could be done. Do you see how to do it?

" Here are two easy ones," said Leonardo, " but each is like one you have already had: How can you show that half of 18 is 10?"

" That takes me back to some of our Roman problems," said Titus.

" Hardly," answered Leonardo. " Here is the other one: What number, not composed of ones, is the same when turned wrong side up?"

" These are too easy," said the boys.

"If you wish a real bad catch, here is one," said Johann. "Take eleven marbles, take away five, add three, and the result is eight."

"That can't be possible," said Leonardo. "The result is nine."

"That depends on how you do it," answered Johann. "I told you it had a bad catch in it."

It must have been a bad one, because it puzzled the rest of the boys.

"I have an old English puzzle," said Cuthbert, "but I know you cannot guess it."

"There is no harm in trying," said Wu.

"Well," said Cuthbert, "show how to take 45 from 45 and have 45 left."

After the others had thought for several minutes, Cuthbert wrote this on a piece of paper and passed it around the room:

$$9 + 8 + 7 + 6 + 5 + 4 + 3 + 2 + 1 = 45$$
$$1 + 2 + 3 + 4 + 5 + 6 + 7 + 8 + 9 = 45$$
$$8 + 6 + 4 + 1 + 9 + 7 + 5 + 3 + 2 = 45$$

He then said: "If you subtract just as with whole numbers, you will find the difference as I have written it. That is, $11 - 9 = 2$, $11 - 8 = 3$, $12 - 7 = 5$, and so on."

"It seems to me," said Robert, "that these are good problems for us to sleep on. We have had an interesting evening, and have heard some curious problems, and now it is nearly twelve o'clock."

"It can't be," said Chang.

"It must be nearly so," said Adriaen.

"You must be wrong," said Wu. "It is twelve now, and — "

But before he could say another word, way out in the dark forest a bell began to toll the midnight hour; and where there was a company of boys about the great roaring fire, there was now a roomful of empty chairs; and where there were a half-dozen burning logs, there was now only a mass of ashes.

The owl looked into the window and blinked wisely. The squirrel scratched his head in a puzzled fashion. The rabbit seemed listening to his own heart beating. The forest was asleep, and the log cabin was silent except for a few lonesome puzzles which were left to cover themselves with blankets of leaves and lie down to sleep on the warm hearthstone in anticipation of other nights with the Memories.

130

THE QUESTION BOX

1. What is the answer to Mohammed's problem about the Arabs?

2. What is the answer to the problem about the chain?

3. What is the answer to the problem about the eggs?

4. How long would it take you to cut a 60-yard piece of cloth into 1-yard lengths, at 1 minute to each cut?

5. Show that half of 88 is twice as much as nothing and that half of 888 is three times as much.

6. If you walk halfway to the door and then walk half of the remaining distance, and then half of what is left, and so on continually, how long will it take you to get out of the room?

7. A man with $1 wanted $1.25. He pawned the $1 for 75 cents and then sold the pawn ticket for 50 cents. He then had his $1.25. Who lost on the transaction?

8. In a certain town three per cent of the inhabitants are one-legged and half of the others go barefoot. How many shoes are necessary?

PRONOUNCING VOCABULARY

Abacus, ăb'à kŭs

Achilles, à kĭl'ēz

Acropolis, à crŏp'ŏ lĭs

Adriaen, äd'rē än

Ahmes, ä'mĕs

Alexandria, ăl'ĕḡ zăn'drĭ à

Alpha, ăl'fà

An-am, ä näm

Athens, ăth'ĕnz

Babylon, băb'ĭ lŏn

Babylonian, băb'ĭ lō'nĭ ăn

Bagdad, bàg däd'

Bel, bĕl

Beta, bā'tà

Bruges, brōō'jĕz

Caius, kā'yŭs

Calandri, kä län'drē

Calculi, kăl'kŭ lē

Calculus, kăl'kŭ lŭs

Caliph, kā'lĭf

Chang, chäng

Cuneiform, kŭ nē'ĭ fôrm

Cuthbert, kŭth'bērt

Daniel, dăn'yĕl

Egypt, ē'jĭpt

Euphrates, ů frā'tēz

Filippo, fĕ lēp'pō

Gerbert, zhĕr bâr'

Gupta, gōōp'tä

Harun-al-Rashid, hä rōōn' är rà shēd'

Heron, hē'rŏn

Hippias, hĭp'ĭ às

Hwangho, hwänghō'

Indus, ĭn'dŭs

Jakob, yä'kŏp

Johann, yŏ hän'

Korea, kŏ rē'à

Leonardo, lā ŏ när'dŏ

Lugal, lōō'gäl

Menes, mē'nēz

Mesopotamia, mĕs'ŏ pŏ tā'mĭ à

Michael, mī'kå ĕl

Mohammed, mŏ hăm'ĕd

Nile, nīl

Papyrus, pà pī'rŭs

Pergamon, pûr'gà mŏn

Pisa, pē'zä

133

Record, rĕk'ŏrd

Seki, sā kē

Shantung, shän'tŏŏng'

Soroban, sȯ'rō bän'

Stevin, stā vȧn'

Stifel, stē'fĕl

Suan pan, swän pän

Sylvester, sĭl vĕs'tẽr

Tigris, tī'grĭs

Titus, tī'tŭs

Tonstall, tŭn'stăl

Van Roomen, văn rō'mĕn

Widman, vĭd'män

Wu, wōō

Yih King, yē kĭng

Yu, yōō

Merchant Books

INDEX

NUMBER PUZZLES

BEFORE THE LOG FIRE

BEING THOSE GIVEN IN THE

NUMBER STORIES OF LONG AGO

BY

DAVID EUGENE SMITH

Merchant Books

NOTE TO TEACHERS

The puzzles which appear in Chapters IX and X of " Number Stories " are here reproduced, but for evident reasons the precise wording has not been retained.

It is hardly necessary to say that the puzzles are not original. There is a considerable literature on the subject, extending back to the Greek mathematicians, and from this historical material selections have been made.

It is proper in this connection to add a few words on certain points that arise in connection with "Number Stories." As stated in the preface to that work, the historical facts were set forth as precisely as circumstances allow. In a case like Robert Record, for example, that writer also spelled his name " Recorde," but in a work for children it was thought best to take the simpler form, since he used this as well as the other. Similarly, the name Caius was chosen because it is a common form, although from the linguistic standpoint it might more properly appear as Gaius, a spelling that is much less familiar to general readers. With respect to such forms as the Egyptian, Babylonian, Chinese, and Roman numerals the attempt was simply made to give a general impression of various methods of writing numbers. If the work had been written for scholars it would have been necessary to consider the use of inverted characters, of numerous other early forms, and of various other sequences of symbols.

NUMBER PUZZLES BEFORE THE LOG FIRE

Page 99. If you take any number, say 3827, and simply mix the figures up, as in 2783, and then take the difference, this difference is always divisible by 9.

The reason is that the remainder arising from dividing any number by 9 is the same as the remainder arising from dividing the sum of its digits by 9. Therefore we may change the order all we wish, and the subtracting of one number from the other will just cancel this remainder.

Page 100. If you write any number of four figures, I will write a number underneath and you will find that the sum of the two numbers is 9999.

This I do by simply subtracting each digit from 9; that is, if you write 4726, I write 5 (which is 9 — 4), 2 (which is 9 — 7), 7 (which is 9 — 2), and 3 (which is 9 — 6), my number being 5273 and the sum being 9999. If you write two numbers, I write two in the same way that I wrote 5273. The sum will then be 2 × 9999, or 19,998. In the same way, if you write three numbers, I will write three numbers, and the sum will be 3 × 9999, or 30,000 — 3, or 29,997, and so on.

Page 101. What figure must be annexed to 74,289 in order to have the number divisible by 11?

The answer is 6, the number then becoming 742,896.

This is found by the help of a law in arithmetic which says that a number is divisible by 11 if the difference between the sums of the digits in the odd and even places is 0 or any other multiple of 11. In the case of 74,289 we see that the sum of the digits in the odd places (9 + 2 + 7) is 18, while the sum of those in the even places (8 + 4) is 12. Taking the difference, 18 − 12 = 6, so I must annex 6 in order to have 18 − 18 = 0.

The number 742,896 is also divisible by 9 because the sum of its digits (7 + 4 + 2 + 8 + 9 + 6 = 36) is divisible by 9. Since it is divisible by 9, it is also divisible by 3. Since it ends in an even number, it is also divisible by 2. Because it is divisible by 2, 3, 9, and 11, it is also divisible by such products as 2 × 3, 2 × 9, 2 × 11, 3 × 9, 3 × 11, and so on.

Page 102. Think of a number, multiply it by 2, add 18, divide the result by 2, subtract the original number, and the result is always 9.

The explanation becomes very simple after a pupil knows a little algebra, for

$$\frac{2x + 18}{2} - x = 9$$

whatever the value of x. Although the explanation may not be adapted to the pupils, the puzzle is none the less interesting.

Page 103. Think of a number, multiply it by 6, add 12, divide by 3, subtract 2, divide by 2, subtract the original number, add 9, and the result is always 10.

The explanation is easily understood by algebra, for

$$\left(\frac{6x + 12}{3} - 2\right) \div 2 - x + 9 = \frac{2x + 4 - 2}{2} - x + 9$$
$$= x + 1 - x + 9$$
$$= 10$$

whatever the value of x.

Which is correct: 6 and 7 *are* 14, or 6 and 7 *is* 14?

Neither, for 6 + 7 = 13, but either "is" or "are" is correct.

2

If 6 cats eat 6 rats in 6 minutes, how many cats will it take to eat 100 rats in 100 minutes, at the same rate?

It will take 6 cats. For

$$6 \text{ cats eat } 6 \text{ rats in } 6 \text{ minutes};$$

so that   1 cat eats 6 rats in 36 minutes,

         1 cat eats 1 rat in 6 minutes,

         1 cat eats 100 rats in 600 minutes,

         6 cats eat 100 rats in 100 minutes.

Page 104. Explain the following:

Ten fingers have I on each hand

Five and twenty on hands and feet.

Simply punctuate the couplet as follows:

Ten fingers have I; on each hand

Five; and twenty on hands and feet.

Write an even number, using only odd digits.

There are many ways, such as $3\frac{3}{3}$, $5\frac{5}{5}$, and so on.

Write a hundred without using zero.

It can be done in various ways; for example, $99\frac{9}{9}$, C, hundred.

Page 105. Find a simple way of writing the products 18×37, 21×37, 24×37, and 27×37.

From the work given on page 105 of "Number Stories," we see that

$$18 \times 37 = 666, \quad \text{and} \quad 6 + 6 + 6 = 18$$
$$21 \times 37 = 777, \quad \text{and} \quad 7 + 7 + 7 = 21$$
$$24 \times 37 = 888, \quad \text{and} \quad 8 + 8 + 8 = 24$$
$$27 \times 37 = 999, \quad \text{and} \quad 9 + 9 + 9 = 27$$

3

Page 106. Find a simple way of writing the products 28 × 15,873, 35 × 15,873, 42 × 15,873, and 49 × 15,873.

Noticing that 28 = 4 × 7, 35 = 5 × 7, and so on, and referring to the results on page 105, we see that

$$28 \times 15,873 = 444,444$$
$$35 \times 15,873 = 555,555$$
$$42 \times 15,873 = 666,666$$
$$49 \times 15,873 = 777,777$$

From this we see that we must multiply 15,873 by 8 × 7, or 56, to obtain 888,888, and by 9 × 7, or 63, to obtain 999,999.

Page 107. Write twelve thousand twelve hundred and twelve.

Evidently we may write these numbers separately. If we add them we have 13,212, but of course no one would read this "twelve thousand twelve hundred and twelve." We have the same trouble in writing eleven thousand eleven hundred and eleven.

$$
\begin{array}{r}
12,000 \\
1,200 \\
12 \\
\hline
13,212
\end{array}
$$

In arithmetic the word "and" is usually omitted in cases like the above.

Page 108. Which is greater, six dozen dozen or a half dozen dozen?

Six dozen dozen is 6 × 12 × 12, while a half dozen dozen is only 6 × 12.

Take nine from six, ten from nine, and fifty from forty, and show that the remainder is six.

$$
\frac{SIX}{IX}{S} \; + \; \frac{IX}{X}{I} \; + \; \frac{XL}{L}{X} \; = \; SIX
$$

4

Page 110. Write one hundred, using only the nine digits and the signs of arithmetic.

There are various ways besides the one given on page 110. For example,

$$74 + 25 + \tfrac{3}{8} + 1\tfrac{9}{8} = 100$$
$$95 + 4 + \tfrac{38}{76} + \cdot\tfrac{1}{2} = 100$$
$$98 + 1 + \tfrac{3}{8} + \tfrac{27}{54} = 100$$

Page 111. Find a number of two digits which is equal to twice the product of those digits.

The number is 36, for

$$36 = 2 \times 3 \times 6.$$

A snail crawling up a pole 10 feet high climbs 3 feet every day and slips back 2 feet every night. How long will it take to reach the top?

Eight days. At the beginning of the second day it is 1 foot above the ground; at the beginning of the third day, 2 feet; and so on, to the beginning of the eighth day, when it is 7 feet above the ground. But on the eighth day the snail crawls up 3 feet more and so gets to the top.

Write figures in place of the letters given below so as to make the multiplications correct:

I C C	I N U
I N	N U
N T T	L N U
I C C	N U S
I A N T	O I N U

Simply take the following key word:

$$\text{I N O C U L A T E S}$$
$$1\ 2\ 3\ 4\ 5\ 6\ 7\ 8\ 9\ 0$$

Page 112. Write figures in place of the letters given below so as to make the divisions correct:

```
              E M A                        M T
     M A ) U E M A          E M A ) U U S S
         M A                        M A S
       ─────                      ─────
         T M                        O S S
         A S                        I A S
       ─────                      ─────
         E M A                      A S
         E M A
       ─────
```

Simply take the following key word:

E M U L A T I O N S
1 2 3 4 5 6 7 8 9 0

A watermelon weighs ⅔ of its weight and ⅔ of a pound. How many pounds does it weigh?

It weighs 4 pounds, for ⅗ of 4 pounds + ⅗ of a pound

$$= \tfrac{4}{5} \text{ of } 5 \text{ pounds}$$
$$= 4 \text{ pounds.}$$

If a herring and a half costs a cent and a half, how much will a dozen and a half herrings cost?

1½ doz. herrings will cost 1½ doz. cents, or 18 cents.

Write nineteen, using only four 9's.

$$9 + 9 + \tfrac{9}{9} = 19.$$

Write two, using only four 9's.

$$\tfrac{9}{9} + \tfrac{9}{9} = 2.$$

6

Page 113. Write twenty, using only four 9's.

$$9 + \tfrac{9\,9}{9} = 20.$$

A bottle and a cork cost $1.10, and the bottle costs $1 more than the cork. How much does each cost?

The bottle costs $1.05, and the cork costs 5 cents.

Which is the heavier, a pound of gold or a pound of feathers?

A pound of feathers is the heavier, because a pound of gold would be weighed by Troy weight, in which the pound is lighter than in the avoirdupois weight. A pound of lead and a pound of feathers weigh the same. A Troy pound is 5760 grains, while an avoirdupois pound is 7000 grains.

How many minutes will it take to cut a strip of cloth 50 yards long into strips 1 yard long if each cut takes one minute?

It will take 49 minutes to cut off 49 yards, and the remaining yard, or fiftieth yard, does not need cutting.

After cutting off $\frac{1}{10}$ of a piece of cloth, a merchant had 100 yards left. How many yards did he have at first?

He had $111\frac{1}{9}$ yards at first.

Page 114. How many quarter-inch squares does it take to make an inch square?

It takes 4 × 4, or 16.

How many quarter-inch cubes does it take to make an inch cube?

It takes 4 × 4 × 4, or 64.

If an apple just balances in weight with ¾ of an apple of the same weight and ¾ of an ounce, how much does the apple weigh?

The apple weighs 3 ounces, for 3 ounces is equal to ¾ of 3 ounces + ¾ of an ounce.

Two fathers and two sons divided three apples among themselves, each receiving exactly one apple. How was this possible?

Suppose that the persons were Mr. Jones, his son, and his grandson. There are two fathers in the group and also two sons, but there are only three persons.

Page 115. In a family party there were 1 grandfather, 2 fathers, 1 grandmother, 2 mothers, 4 children, 3 grand-children, 1 brother, 2 sisters, 2 sons, 2 daughters, 2 married men, 2 married women, 1 father-in-law, 1 mother-in-law, and 1 daughter-in-law. How many were there in the party?

There might have been only seven — an old man and his wife together with their son and his wife, his two daughters, and his son. If these are counted, it will be seen that there are 1 grandfather, 1 grandmother, 2 fathers, and so on, as the problem states.

Page 116. Find a number which when multiplied by 3 and the result added to 8 becomes 35.

The number is 9, as is seen by working backward from 35. In algebra it is found by solving the equation $3x + 8 = 35$.

Explain this old puzzle:

> Every lady in this land
> Has twenty nails upon each hand
> Five and twenty on hands and feet
> And this is true without deceit.

8

Simply punctuate as follows:

> Every lady in this land
> Has twenty nails; upon each hand
> Five; and twenty on hands and feet.

Write 78, using only the figure 7, but repeating it as often as you wish.

Write it like this: $77\frac{7}{7}$, or $77\frac{77}{77}$, and so on.

The rest of the puzzles in the Question Box are sufficiently evident from the puzzles given in the text.

Page 119. There were two Arabs who sat down to eat, one with five loaves, and the other with three, all the loaves having the same value. Just as they were about to begin, a third Arab came along and proposed to eat with them, promising to pay eight cents for his part of the meal. If they ate equally and consumed all the bread, how should the eight cents be divided? Remember that one Arab furnished five loaves and the other furnished three loaves.

One should have 7 cents, and the other 1 cent.

Each Arab ate $\frac{1}{3}$ of 8 loaves, or $2\frac{2}{3}$ loaves. The Arab who furnished 5 loaves gave to the visiting Arab $5 - 2\frac{2}{3}$ loaves, or $2\frac{1}{3}$ loaves; and $2\frac{1}{3} = \frac{7}{3}$. The other Arab, who furnished 3 loaves, gave to the visiting Arab $3 - 2\frac{2}{3}$ loaves, or $\frac{1}{3}$ of a loaf. So one Arab gave 7 times as much as the other and should receive 7 times as much money.

A man had five pieces of chain, each having three links. He asked a blacksmith how much he would charge to make them into one piece of chain. The blacksmith replied that he charged two cents to cut a link and two cents to weld or fasten together a link. How much was the blacksmith's charge?

It would cost 12 cents. Simply cut each of the three links of the last part and fit these links between the other parts. You then cut three links and weld three links.

Page 120. How may you arrange the figures 1, 6, 8 so that the three-figure number that you make shall be exactly divisible by 9?

In the number 168 turn the 6 wrong side up, making 198.

When chess was invented, the Persian king was filled with great joy and commanded that chessboards should be placed in all the temples of the kingdom. "Moreover," he said to the inventor of the game, "ask what you will." Thereupon the inventor replied, "Give me, O king, a single grain of wheat to place in the first square of the chessboard, two in the second, four in the third, and so on, doubling the number each time, until all the sixty-four squares are filled." How many grains were needed?

The result could be found by adding 1, 2, 4, 8, and so on, until 64 numbers were used. Algebra tells us a shorter method:

$$2^{64} - 1 = 18,446,744,073,709,551,615$$

Page 121. A man wanted a blacksmith to shoe his horse. The blacksmith said he would do so if he were paid one cent for the first nail, two cents for the second, four cents for the third, eight cents for the fourth, and so on, doubling the amount each time. There being twenty-four nails, how much did it cost to shoe the horse?

The result may be found by adding. Algebra tells us a shorter method:

$$2^{24} - 1 = 16,777,215$$

The amount in dollars is $167,772.15.

Page 122. A girl was carrying a basket of eggs, and a man driving a horse hit the basket and broke all the eggs. Wishing to pay for the damage, he asked the girl how many eggs she had. The girl said she did not know, but she remembered that when she counted them by twos, there was one egg left over; when she counted them by threes, there was also one egg left over; when she counted them by fours, there was also one left over; but when she counted them by fives, there were no eggs left over. How many eggs were there?

She had 25 eggs. We can easily see this, because the number is divisible by 5 and hence must be in the series 5, 10, 15, 20, 25, 30, and so on. We have simply to select a number out of this series which has a remainder of 1 when divided by 2, 3, or 4. The only number of reasonable size that meets these requirements is 25. _____

Page 123. A mule and a horse were carrying some bales of cloth. The mule said to the horse, "If you give me one of your bales, I shall carry as many as you." "If you give me one of yours," replied the horse, "I shall carry twice as many as you." How many bales was each carrying?

The mule carried 5 bales and the horse carried 7.

At one time the clocks of Venice struck from 1 to 24. How many strokes were there in twenty-four hours?

We might find the result by adding, thus: $1 + 2 + 3 + 4 +$, and so on to $+ 23 + 24$. Algebra furnishes us with a shorter method:

$$\tfrac{1}{2} \text{ of } (1 + 24) \times 24 = 25 \times 12 = 300$$

_____

Page 124. Suppose that Achilles can run ten times as fast as a turtle and that the turtle has a mile the start. How long will it take Achilles to overtake the turtle?

In 10 minutes Achilles will go 1 mile, and the turtle will go $\frac{1}{10}$ of a mile, so that in 10 minutes Achilles will gain $\frac{9}{10}$ of a mile on the turtle. If he gains $\frac{9}{10}$ of a mile in 10 minutes, he will gain $\frac{9}{100}$ of a mile in 1 minute. To gain 1 mile it will take him $1 \div \frac{9}{100}$ minutes, or $\frac{100}{9}$ of a minute, or $11\frac{1}{9}$ minutes.

This is easily seen by computing the distance that each will go in $11\frac{1}{9}$ minutes. ————————

Page 126. Show how to distribute ten pieces of sugar in three cups so that every cup shall contain an odd number of pieces.

Place 7 pieces in one cup and 3 in another, and then place one of these cups in the third cup.

In a box there are six apples. Divide these equally among six boys in such a way as to leave one apple in the box, but do not cut any apple.

Give one apple to each of 5 boys and then give the other boy the box with the remaining apple in it.

A man sold his farm for $5000, which is what it cost him, then bought it back for $4500, and then sold it again for $5500. How much did he gain?

He gained $1000. He began with a $5000 farm, and his expenditures were $5000 (farm) + $4500, while his receipts were $5000 (farm) + $5500. His total expenditures were therefore $9500 and his total receipts $10,500.

————————

Page 127. What is the number of two figures which multiplied by 8 becomes 20?

The answer is $\frac{5}{2}$.

Show that six and six are eleven.

Take a Roman VI and turn it upside down, making ΛI. On top of this put VI and you have XI.

Show that four and four are nine.

Take a Roman IV and turn it upside down, making IΛ. On top of this put IV and you have IX.

Three brothers divided four apples among themselves so that one had no more than the others, and yet no apple was divided. Explain how this was possible.

Give one boy two apples and each of the others one apple. Then no boy has no more than the *others*, although one of them has more than *either* of the others.

———

Page 128. Show that half of 18 is 10.

Simply draw a horizontal line through 18 so that each half will be 10.

What number is the same when turned wrong side up?

Simply turn 69 through 180° without lifting it from the paper, and it becomes 69 again. We might take 96, 6969, 9696, and so on.

———

Page 129. Take eleven marbles, take away five, add three, and the result is eight.

Simply add the 3 to the 5 and the result is 8.

———

Page 131. How long would it take you to cut a 60-yard piece of cloth into 1-yard lengths, at 1 minute to each cut?

It would take 59 minutes. The sixtieth piece would not need to be cut.

Show that half of 88 is twice as much as nothing.

Draw a horizontal line through the number and each half will be 00.

If you walk halfway to the door and then walk half of the remaining distance, and then half of what is left, and so on continually, how long will it take you to get out of the room?

It would be impossible ever to get out of the room, however fast you walk. If, however, it takes 1 minute to walk the first distance, $\frac{1}{2}$ minute to walk the next distance, and so on, the number of minutes taken to reach the door is $1 + \frac{1}{2} + \frac{1}{4} + \frac{1}{8} +$, and so on. Algebra teaches us that the sum of these numbers cannot exceed 2, so that you could reach the door in 2 minutes.

A man with $1 wanted $1.25. He pawned the $1 for 75 cents and then sold the pawn ticket for 50 cents. He then had his $1.25. Who lost on the transaction?

The man who bought the pawn ticket, since it was good for only 25 cents, the difference between $1 and 75 cents.

In a certain town three per cent of the inhabitants are one-legged and half of the others go barefoot. How many shoes are necessary?

As many shoes as there are inhabitants.

The one-legged group each wear one shoe. Of the remainder of the population, if one half go barefoot, that is equivalent to all the remaining population wearing one shoe apiece; therefore it will take the equivalent of one shoe to every person for the town to be shod in accordance with the terms of the problem. The three per cent might as well be any other per cent, and it makes no difference what the population is. The answer always is as many shoes as there are people in the town.

The rest of the puzzles in the Question Box are sufficiently evident from the puzzles given in the text.

Lightning Source UK Ltd.
Milton Keynes UK
UKHW032056241122
412811UK00007B/146